ACRYLIC POLYMERS IN HEALTHCARE

Edited by **Boreddy S.R. Reddy**

Acrylic Polymers in Healthcare
http://dx.doi.org/10.5772/66610
Edited by Boreddy S.R. Reddy

Contributors

Lavinia Cosmina Ardelean, Cristina Maria Bortun, Laura Cristina Rusu, Angela Podariu, Oluwafemi Ayodeji Adebo, Eugenie Kayitesi, Janet Adebiyi, Sefater Gbashi, Makumba Temba, Adeseye Lasekan, Judith Phoku, Patrick Berka Njobeh, Ángel Serrano-Aroca, Ma. Concepción Arenas-Arrocena, Laura Susana Acosta-Torres, Liliana Argueta-Figueroa, René García-Contreras, Omar Martinez, Javier De La Fuente, Berenice Alejandra Camacho-Flores, Pilar Rodriguez-Torres, GAGANDEEP Kaur, Brijesh Kumar, S.B. Rai, Velichka Andonova, Bozena Tyliszczak, Anna Drabczyk, Sonia Kudlacik, Aslı Erdem Yayayürük

Published by InTech

Janeza Trdine 9, 51000 Rijeka, Croatia

Publishing Process Manager Nina Kalinic
Technical Editor SPi Global
Cover InTech Design team

First published October, 2017
Printed in Croatia
Legal deposit, Croatia: National and University Library in Zagreb

Additional hard copies can be obtained from orders@intechopen.com

Acrylic Polymers in Healthcare, Edited by Boreddy S.R. Reddy
p. cm.
Print ISBN 978-953-51-3593-7
Online ISBN 978-953-51-3594-4

For EU product safety concerns:
IN TECH d.o.o., Prolaz Marije Krucifikse Kozulić 3, 51000 Rijeka, Croatia,
info@intechopen.com or visit our website at intechopen.com.

Meet the editor

Dr. Boreddy S.R. Reddy is currently settled in Watford, Hertfordshire, England (UK), after retirement as the director of the Grade Scientist (G) and the head of the Industrial Chemistry in the World-Renowned Institute for Science, Research and Leather Technology under the umbrella of CSIR, Government of India, namely, Central Leather Research Institute, Chennai, Tamil Nadu (India), for the last 28 years. He received his BSc degree from the Hindu College, Machilipatnam, Andhra University (India); MSc degree from the Sri Venkateswara University Post Graduate Centre, Anantapur (India); and PhD (1978) and DSc (2005) degrees from the University of Madras, Chennai (India).

Prof. B.S.R. Reddy's main areas of research during his 38 years of varied experience in research and teaching both in the UK and India include applied synthetic organic chemistry and polymer science/technology, ring-opening metathesis, synthesis and polymerization of a dictionary of several functionally activated acrylates, functional microspheres, polymer drug conjugates, polymer supports, electroactive polymers, P–F reactions leading to synthetic tanning agents and adhesives, (nano) composites, organic-inorganic composites and functional polymers. He has published 250 research papers as author/co-author in leading international journals of reputation and has 12 patents to his credit including editing 2 books by contributing some chapters. To be precise, he is one of the most frequently cited authors in chemical, pharmaceutical and biological sciences.

Prof. Reddy's academic achievements were duly recognised by the leading professional bodies by electing as fellow of the Royal Society of Chemistry (London) and the Society of Polymer Science (India). Also, he was elected as fellow of the Andhra Pradesh Akademi of Sciences (India) and the Tamil Nadu Academy of Sciences (India). The Government of Tamil Nadu honoured Dr. B.S.R. Reddy by awarding a prestigious Tamil Nadu Scientist Award (TANSA) in 1997 for his wide-ranging contributions to science and society.

Contents

Preface

Polymer chemistry occupied a central position among the sciences, having a close association with physics via the concepts and principles of physical chemistry and with biology through its application to biological aspects. The chemical technology, especially the branch of acrylic polymers in polymer chemistry and technology, has contributed fundamentally towards the development of healthcare in recent times. The development and the application of acrylic polymers towards dentistry and drug delivery systems are dependent on the chemistry to provide explanations necessary for their development.

The combination of polymer composition, chain length, branching, cross-linking and molecular orientation can produce a variety of requisite properties. Towards the needs of various dental applications and drug delivery systems, these features are manipulated to produce a balance of optimum properties that approach the ideal performance properties as closely as is practical. The properties are grouped into four interrelated categories, namely, mechanical, thermal, rheological (flow) and dissolution in a relevant medium. The applied forces produce stresses within polymers for prosthetics and drug delivery applications that cause materials to deform (undergo strain) via either elastic or plastic or a combination of elastic and plastic strain (viscoelastic strain).

The other major contributing factor is cross-linking by using materials having hydro-/lipophilic units. In the absence of cross-linking, only relative weak inter-polymer-chain bonds (van der Waals and hydrogen bonds) are available to hold the polymer chains together in a solid state. The chain slippage decreases as the chain length increases because the bonds between chains, together with chain entanglements, resist dislodgement of the individual chains. At a definite chain length, the resistance provided by the inter-chain bonds and entanglements becomes strong enough to exceed the covalent strength of the C–C bonds along the backbone chains. At this critical chain length, an applied force can rupture chains rather than dislodge them and cause one chain to slide past another. The balance between the strengths of the interaction bonds and the covalent bonds along the backbone chains explains why the physical and mechanical properties of polymers increase with increased molecular weight up to a certain point.

This book presents eight chapters organised under three parts vividly by providing a wealth of new ideas in design, synthesis and a detailed study of new acrylate materials in healthcare applications. Most chapters are introduced with some relevant features. This helps to focus attention on some of the more important details and provides a framework upon which learning and understanding can be built. These also help to identify specific points that researchers frequently attempt to incorporate when conducting an advanced research.

Part I represents Chapters 1, 2, 3 and 4. Chapter 1 deals with acrylates and their alternatives in dental applications. Here, three aspects were covered, namely, mechanical properties (fracture behaviour, water absorption and mechanical strength degradation) caused by the exposure to saliva of classical heat-cured acrylic resins, studies concerning biocompatibility (in vivo and in vitro tests) of acrylic resins and alternative resins and technologies for denture manufacturing. Chapter 2 describes about the acrylates in dental applications. Here, the role played by acrylates in dentistry has been highlighted. The composition, functions and types of teeth have been described. The materials including dentures, adhesives, impression trays and dental crowns are used widely in dentistry and dental prosthetics in the preparation of which the key role is played by acrylates have been characterised. Special attention has been drawn on the possibility of modification of the synthesis of acrylic materials that can lead to the improvement of their properties and result in making them more favourable from the point of view of the patient. Chapter 3 represents some new trends for the processing of poly(methyl methacrylate) (PMMA) biomaterial used for dental prosthodontics. The main aspects covered here are the artefacts of synthesis, morphology and the biological properties, namely, cytotoxicity, mutagenicity, biocompatibility and the anti-microbiological activity of PMMA. Chapter 4 carried out the latest improvements of acrylic-based polymer properties for biomedical applications. The acrylics have been fabricated as multicomponent polymeric systems in the form of interpenetrated polymer networks or combined with other advanced materials such as fibres, nanofibres, graphene and its derivatives and/or many other kinds of nanoparticles (NPs) to form composite or nanocomposite materials, which are expected to exhibit superior properties. Besides, in regenerative medicine, acrylic scaffolds are designed with the required extent, and morphology of pores is characterised by sophisticated techniques.

Part II comprises Chapters 5 and 6. In Chapter 5, synthetic polymer-based nanoparticles as intelligent drug delivery systems are described. This chapter aims to look at the 'hot spots' in the design of synthetic polymer nanoparticles as an intelligent drug delivery system in terms of biopharmaceutical challenges and in relation to the route of their administration, namely, the non-invasive—oral, transdermal, transmucosal (nasal, buccal/sublingual, vaginal, rectal and ocular) and inhalation routes—and the invasive parenteral route. Chapter 6 presents spectroscopic investigations of poly(vinyl alcohol) film with complex of terbium ions along with bismuth nanoparticles for improved green emission. The bismuth nanoparticles (NPs) have been synthesised by the pulsed laser ablation technique at different pH in different aqueous solutions [viz. water (H), water + sodium hydroxide (HN) and water + hydrochloric acid (HC)]. The NPs in aqueous solutions have been characterised by TEM and UV-Vis-NIR absorption techniques. Then, photoluminescence properties of Tb^{3+} ions and the $[Tb(Sal)_3(Phen)]$ complex were studied using 266 nm and 355 nm as excitation wavelengths to seek into the influence of Bi NPs on their emissive properties.

Part III deals with Chapters 7 and 8. Chapter 7 investigates mitigation of acrylamide (ACR) in foods in the context of an African perspective. Acrylamide (ACR) is a possible human carcinogen, with neurotoxic properties. It is a heat-generated food toxicant particularly found in carbohydrate-rich foods. However, information on the extent of ACR occurrence in foods consumed in different parts of Africa is rather too limited. This covers the formation, occurrence and health impact of ACR in foods. It further summarises previous studies looking at ACR reduction and mitigation strategies, especially those that may be applicable in the African continent. Chapter 8 presents the use of acrylic-based polymers in environmental remediation studies. The polymeric adsorbents have received considerable interest for heavy metal removal mainly due to important technological and scientific developments such as easy synthesis at controlled dimensions with variable functional groups, perfect mechanical rigidity, tunable surface chemistry, large surface area, pore size distribution, high uptake values and feasible regeneration under mild conditions. Also, an attempt to present to the readers the widespread investigations of acrylic-based polymeric adsorbents is made so that they can get an idea about the various types and forms of polymeric materials used for the removal of heavy metals from water.

Dr. Boreddy S.R. Reddy
D.Sc., Ph.D., C.Chem., FRSC (London),
Scientist Emeritus and Director Grade Scientist (Formerly),
Hertfordshire,
England, United Kingdom

Section 1

Acrylates in Dentistry

Chapter 1

Acrylates and Their Alternatives in Dental Applications

Lavinia Cosmina Ardelean, Cristina Maria Bortun,
Angela Codruta Podariu and Laura Cristina Rusu

Additional information is available at the end of the chapter

http://dx.doi.org/10.5772/intechopen.69010

Abstract

Acrylic resins dominated dentures technology for several decades. Due to their many disadvantages, new classes of resins, which promise better quality, constantly appear. Mechanical properties of acrylic resins, including fracture behaviour, water absorption and mechanical strength degradation caused by the exposure to saliva of classical heat-cured acrylic resins compared to alternative urethane-based light-cured resins, were carried out. The allergy potential of acrylic resins was evaluated by *in vivo* and *in vitro* tests. New choices of resins, like thermoplastic injected resins, light-cured or milled high-performance polymers, with better properties compared to acrylics, suitable for dental applications are being presented.

Keywords: acrylates in dentistry, alternative resins for dental use, thermoplastic resins, light-cured resins

1. Introduction

In dentistry, non-metallic materials for denture manufacturing have a long tradition [1]. Among the first materials used, wood, ivory and dentin from hippopotamus teeth or even human teeth may be found. These types of dentures were considered a luxury, due to their prohibitive price and only rich people could afford them. Charles Goodyear discovered vulcanized rubber in 1839. This was the premise for manufacturing dentures with rubber base, much more cheaper and accessible to any pocket [2]. Celluloid, which appeared in 1871, was the first artificial polymer competing with rubber. But, this was not able to overcome the drawbacks such as dimensional instability, deformability and problems in processing technology. Resins represented a huge step forward in dentistry and the first heat-cured acrylic was reported in 1936 [3]. First chemical studies regarding diacrylic composite resins-type urethane polymers were also carried out at that time by Otto Bayer in the IG Farben Laboratories in Leverkusen. Acrylics, in fact poly(methyl methacrylate) (PMMA) mixed with methyl methacrylate, dominated denture technology for several decades. There were no competitors in manufacturing denture bases, artificial teeth, orthodontic appliances, single-tooth or provisional restorations or as veneering materials (**Figure 1**).

The toxicity of the residual monomer, the complex wrapping system, difficult processing and poor resistance are some of the disadvantages of these materials. Many new classes of resins/macromolecular compounds which promise better quality came on to the market such as diacrylic, styrene, polycarbonate, epiminic, polyurethane, vinyl, polyamide, acetal and polyglass. Besides classic heat-curing, alternative technologies namely, casting and injection moulding are nowadays available in manufacturing acrylic resins for dental applications. In the case of alternative resins, light-curing or microwave polymerization techniques are also used [4, 5]. Light-curing, as a polymerization method for dental materials, appeared in the 1970s. Initially, ultraviolet light was used. Afterwards, it was replaced by visible radiation (visible spectrum wavelength/electromagnetic waves), the light source being either a halogen bulb or xenon stroboscopic lamps [6, 7]. The classification of resins according to DIN EN ISO 1567 is presented in **Table 1**.

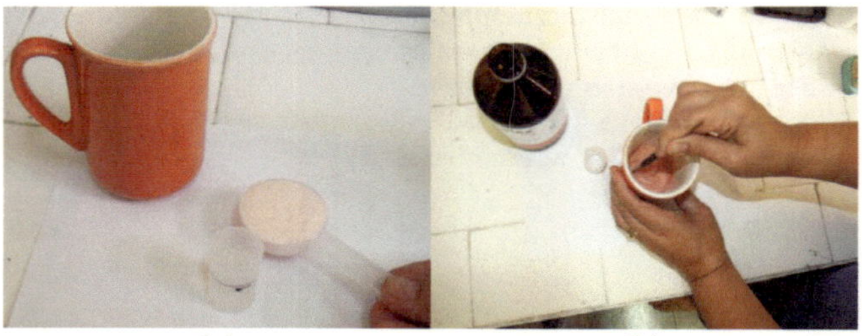

(a) (b)

Figure 1. (a) Heat-curing acrylate powder and liquid and (b) mixing the acrylic paste.

Type	Class (manufacturing)	Group (presentation form)
Type 1	Heat-cured resins (>65°C)	Group 1: Bicomponent powder and liquid Group 2: Monocomponent
Type 2	Self-cured resins (<65°C)	Group 1: Bicomponent powder and liquid Group 2: Bicomponent powder and casting liquid
Type 3	Thermoplastic resins	Monocomponent system grains in cartridges
Type 4	Light-cured resins	Monocomponent system
Type 5	Microwave-cured resins	Bicomponent system

Table 1. The classification of resins according to DIN EN ISO 1567.

2. Mechanical properties of acrylic resins

2.1. Evaluation of water absorption and mechanical strength degradation caused by the exposure to saliva of classical heat-cured acrylic resins compared to alternative urethane-based light-cured resins

It is well known that acrylates for dental use have poor resistance and these degrade in the wet environment of the mouth. Our studies involve evaluation of water absorption and mechanical strength degradation caused by the exposure to saliva of classical heat-cured acrylic resins compared to alternative urethane-based light-cured resins, which are also used for dentures manufacturing. Twenty samples (plates: 2 mm in thickness, 30 mm in length and 5 mm in width) of Meliodent (Heraeus-Kulzer) heat-curing acrylic resin and twenty samples of two urethane-based light-curing resins from the same system-Eclipse Resin System: Eclipse Base Plate and Eclipse Contour Resins (Dentsply-DeguDent) were analyzed, in saliva and dry environment. Ten samples were immersed in saliva with low microbial content and neutral pH, at 37°C for 30 days. The other ten samples were kept dry for 30 days. Saliva was collected from clinically healthy subjects and tested for germs with Vivacare line CRT bacteria 2 in one test kit. The test results showed level two of four possible contaminations and so the saliva was considered not severely contaminated. Its pH, determined with an indicator strip, was normal, with an average value of six, as shown in **Figure 2**. In order to determine the water percentage content, the samples were initially weighed and further weighed after 48, 144, 312 and 720 h.

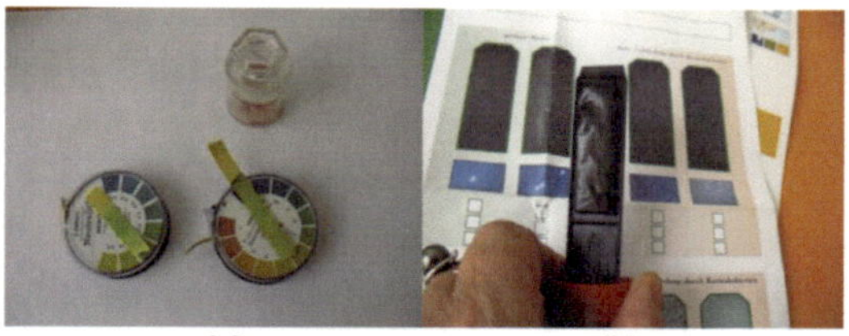

(a) (b)

Figure 2. (a) pH index and (b) quantitative evaluation of Streptococci and Lactobacilli microorganisms.

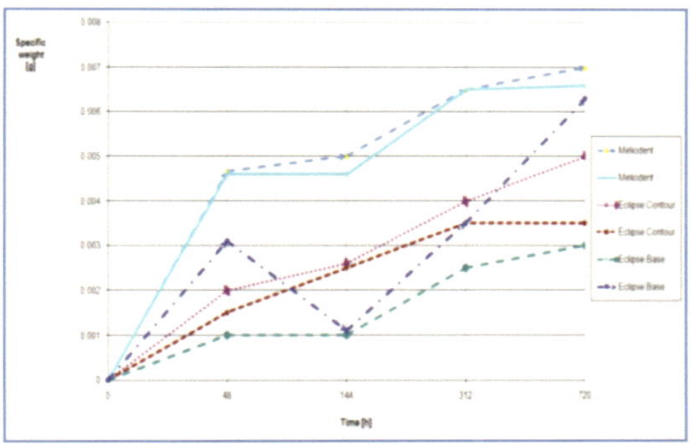

Figure 3. Correlation between specific weight and time for the tested polymers (specific weight = measured weight − initial weight).

Meliodent (Heraeus-Kulzer) was proven to have the highest water absorption capacity, followed by Eclipse Contour Resin and Eclipse Base Plate (Dentsply-DeguDent) (**Figure 3**).

The diagram showing correlation between the three types of resins and the percentage humidity content reveals that the heat-curing resin absorbs more water than the light-curing resins. The diagram indicates the maximal and minimal values for each material (**Figure 4**).

Zwick Roell extensometer (Zwick GmbH & Co.) was used to determine the moment of sample breaking or fracture point and its elongation. TestXpert software was used to standardize the applications (**Figure 5**).

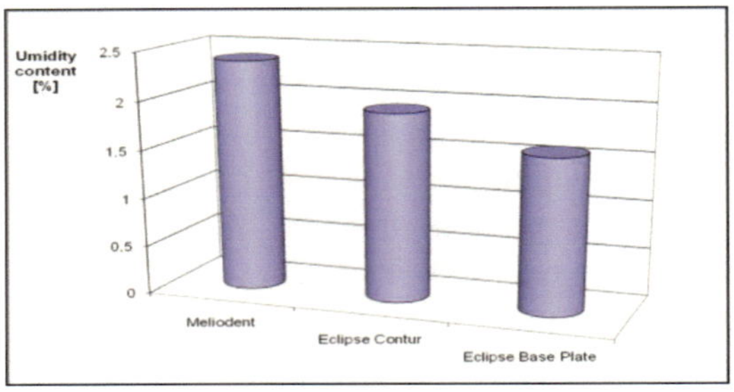

Figure 4. The determination of percentage humidity content of polymer.

Figure 5. The determination of tensile mechanical resistance using the Zwick Roel extensometer.

Materials	Normal conditions		Humid conditions (1 month)	
	E [MPa]	R_m [MPa]	*E* [MPa]	R_m [MPa]
Meliodent	1615.60	52.77	1168.50	48.51
Eclipse base	2527.60	94.46	1921.50	66.76
Eclipse contour	1955.00	40.53	1577.00	25.39

Table 2. The mechanical properties of the studied materials kept in dry and in humid conditions for 1 month.

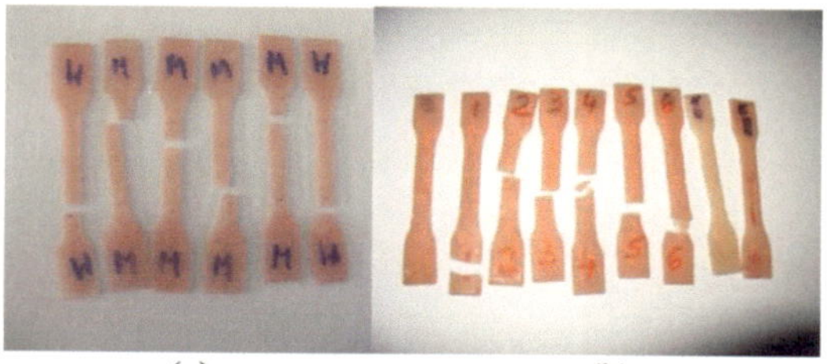

(a) (b)

Figure 6. (a) Meliodent samples broken after stretching and (b) broken Eclipse samples.

Results showed that the humid samples (kept in saliva) had significant lower values than the dry samples, realized from the same material. The Young's elasticity modulus (E) and ultimate tensile strength (R_m) were taken into consideration. The average values after the sample analysis are shown in **Table 2**.

There is an evident difference among the tested materials. The heat-curing resin has a lower value of the elasticity modulus than the urethane-based resins. The values decrease distinctly in humid environment, especially in the case of Eclipse Base resin. The mechanical strength also shows decreased values for the humid samples, for all three resins tested in the present work. Generally, the results indicate a higher decrease of mechanical strength values for urethane-based light-curing resins (Eclipse) when compared to heat-curing acrylate (Meliodent) resins. The heat-curing acrylate shows a higher decrease in elasticity than the light-curing resins (**Figure 6**). The differences noticed between humid and dry environmental conditions indicate that there is clear evident role of the saliva in the biodegradation of the denture base polymers [8].

2.2. Fracture behaviour after elongation of two heat-curing acrylic resins

The same type of samples (plates: 2 mm in thickness, 30 mm in length and 5 mm in width) made of Meliodent (Heraeus-Kulzer) and Royaldent Plus (Palatinal Kft.) heat-curing acrylic resins were tested to compare the fracture behaviour after elongation. Both longitudinal and transversal surfaces of the samples, after breaking, were analyzed using the Olympus type SZX7 stereomicroscope equipped with an image processing system QuickphotoMicro 2.2. soft (**Figure 7**).

The two acrylic resins have different fracture behaviour. In the case of Meliodent resin, the elongation before fracture was lower compared to that of Royaldent resin, indicating a ductile fracture behaviour.

The following data were obtained by testing the two resins using Zwick Roell extensometer (Zwick GmbH & Co.). Meliodent: Ultimate tensile strength (R_m): 54–75 MPa, Young's elasticity modulus (E): 1383–1688 MPa and elongation: 1–3.5%. Royaldent: Ultimate tensile strength (R_m): 69–90 MPa, Young's elasticity modulus (E): 1282–1937 MPa and elongation: 2–5%.

In the longitudinal section, Meliodent samples do not show elongation. The final fracture being sudden compared to the Royaldent samples, which had a significant elongation before fracture. In the transversal section, one may remark that dark reinforcement fibres from Meliodent

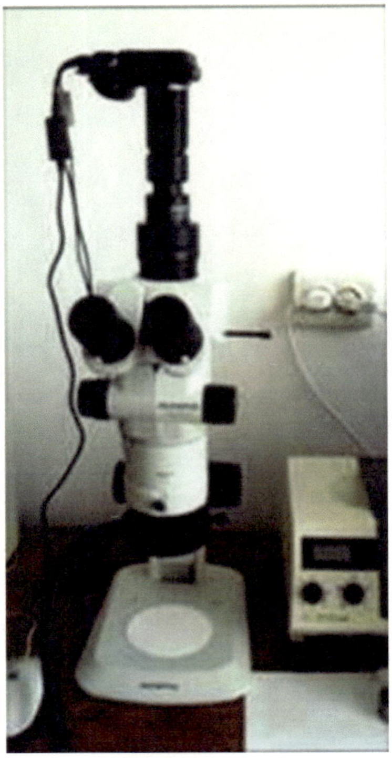

Figure 7. Stereomicroscope Olympus type SZX7.

(**Figure 8a**) do not break together with its polymeric matrix, whereas both matrix and fibres are broken at the same time in the case of Royaldent sample (**Figure 8b**). The different fracture behaviour of the two acrylic resins may be explained by the differences between mechanical characteristics of reinforced fibres and the polymeric matrix. The mechanical characteristics of fibres are better than those of the matrix in the case of Meliodent samples, whereas Royaldent samples showed the similar characteristics for fibres and matrix. Therefore, Royaldent samples show a better behaviour to fracture than Meliodent samples as well as ductile behaviour.

Stereomicroscopic analyses of two acrylic resins showed that the entire sample has a brittle fracture having a quasi-crystalline aspect in the transversal section.

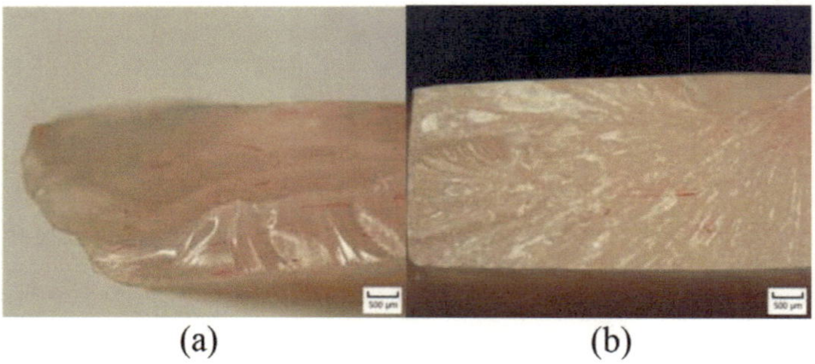

Figure 8. The stereo microstructural aspect of the samples: (a) Meliodent and (b) Royaldent.

2.3. Fracture toughness evaluation

Being long-term prosthetic pieces, complete dentures need a warranty regarding their mechanical resistance and lifetime. The fracture toughness, which reflects its resistance to fracture and represents the energy required for a crack to propagate through a material to its complete fracture, was evaluated in complete denture technology.

Samples without initial cracks were considered for testing so that the value of the stress intensity factor (K_{IC}) depends only on sample's dimension and critical load value.

Two different heat-curing acrylics were selected. Meliodent (Heraeus Kulzer) and Royaldent Plus (Palatinal Kft.), and a light-curing urethane-based resin-Eclipse Resin System (Dentsply-DeguDent) were taken into consideration. The samples were disk-shaped with a circular hole in the centre. Five different samples with the following dimensions were taken for the test:

Sample 1: R_{out} = 42 mm, R_{in} = 4.2 mm, R_{in}/R_{out} = 0.1 and h = 2 mm.

Sample 2: R_{out} = 45 mm, R_{in} = 6.75 mm; R_{in}/R_{out} = 0.15 and h=2 mm.

Sample 3: R_{out} = 48 mm, R_{in} = 9.6 mm; R_{in}/R_{out} = 0.2 and h = 2 mm.

Sample 4: R_{out} = 50 mm, R_{in} = 12.5 mm; R_{in}/R_{out} = 0.25 and h = 2 mm.

Sample 5: R_{out} = 53 mm, R_{in} = 15.9 mm; R_{in}/R_{out} = 0.3 and h = 2 mm.

where R_{in} = hole radius, R_{out} = disk radius and h = disk thickness.

Five samples were made of each material (**Figure 9**).

The compression tests were performed with the same static loading machine (**Figure 6a**), model Zwick Roell of 5 kN (Zwick GmbH & Co.), connected to a computer (TestXpert specific soft.). The samples were compressed until breaking (**Figure 10**).

Each sample was loaded by a pair of point forces, which acted along the diameter. The distribution of the load across the thickness of the disk was uniform. When the force was applied, the micro-cracks situated in the proximity of the force line, at the edge of the inner hole, started to grow, and at a certain value of the force gave rise to a macro-crack. Other

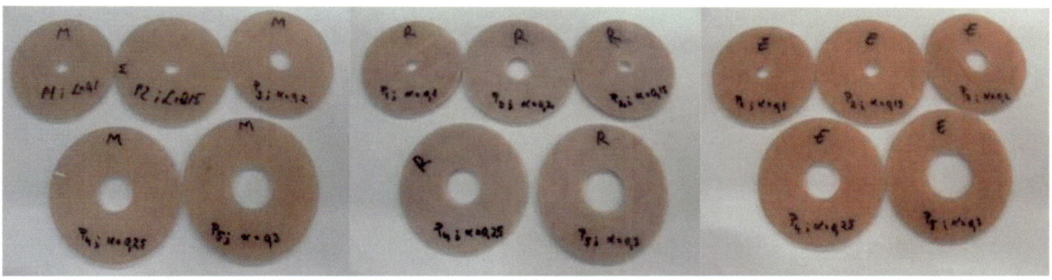

(a) (b) (c)

Figure 9. (a) Meliodent samples, (b) Royaldent samples and (c) Eclipse samples.

Figure 10. Meliodent sample during compression experiment.

pre-existing cracks within the specimen did not grow. Here, the central hole, playing the role of the defect, initiated the fracture. As expected, in general the crack developed symmetrically, beginning from the inner hole as shown in **Figure 11**.

The fracture toughness (K_{IC}) was calculated, depending only on the sample's dimensions and critical value of the load. As the brittle materials have high relative compression values and low tension values, the failure begins at the point of inner boundary. In general, the fracture direction was perpendicular to the loading direction. The following values for fracture toughness were obtained: K_{IC} = 2.4 MPa√m for Meliodent, K_{IC} = 2.65 MPa√m for Royaldent and K_{IC} = 3.35 MPa√m for Eclipse (**Figure 12**).

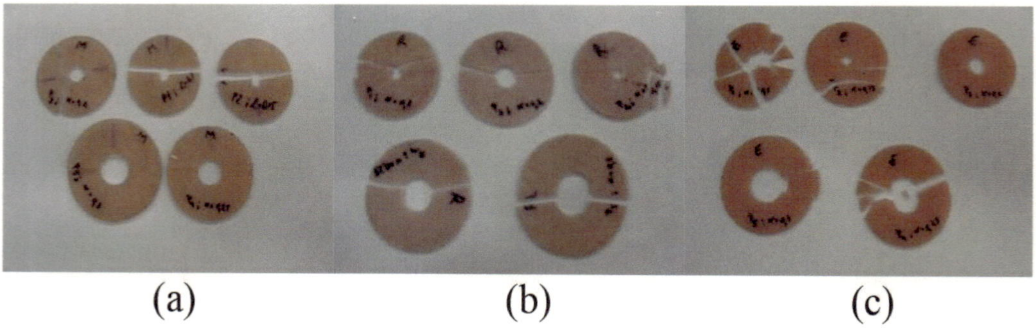

Figure 11. Aspects of the samples after the compression experiment: (a) Meliodent, (b) Royaldent and (c) Eclipse.

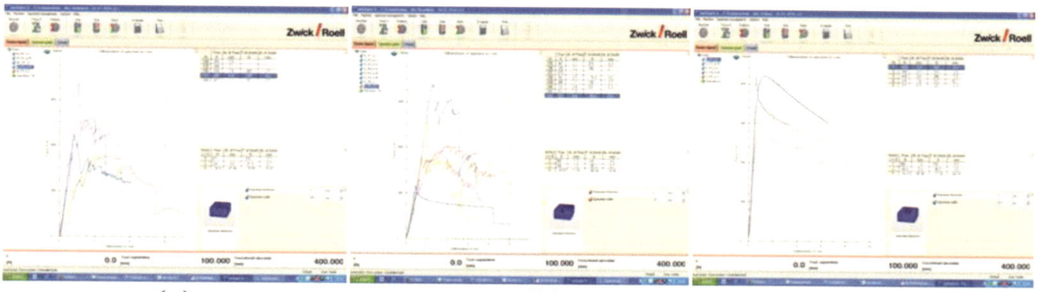

(a) (b) (c)

Figure 12. Force-displacement diagrams resulted after caring out compression test: (a) Meliodent, (b) Royaldent and (c) Eclipse.

The results obtained for the three tested resins showed no significant differences, Eclipse resin shows the highest value for fracture toughness [9, 10].

Comparative studies were undertaken with the same materials using different methods, in order to get comparative results. Single-edge-notched beam (SENB) method and indentation strength (IS) method were used. In the case of a single-edge-notched beam (SENB) method, samples were prepared in the form of plates with dimensions of 50 × 50 × 2 mm from which rectangular beams with dimensions of 4 × 2 × 25 mm (width/thickness/length) were cut.

The bending tests were carried out on a Zwick-Roell 5 kN testing machine (Zwick GmbH & Co.) (**Figure 6a**). The two halves of the broken samples were used for the measurement of the notch depth c. The toughness and the experimental values for *K*ic, obtained using SENB method were 2.26 MPa√m for Meliodent and 3.18 MPa√m for Eclipse resins.

Indentation strength (IS) method uses a Vickers pyramid to determine the fracture toughness by analyzing the stress field at a crack tip. The indentations of the samples were made using a Vickers hardness tester, model HMO 10, in the middle of the tensile surface of the beams at a load of 98 N, for 15 s, magnitude which prevented radial cracks (**Figure 13**).

(a) (b)

Figure 13. (a) The Vickers pyramid and (b) the samples.

Material	Hardness (H) [MPa]	Fracture toughness (K_{IC}) [MPa√m] SENB	Fracture toughness (K_{IC}) [MPa√m] IS
Meliodent	29.54	2.26	2.31 (v = 0.05 mm/min)
Eclipse	25.35	3.18	3.26 (v = 0.05 mm/min)

Table 3. Measurement results for Meliodent and Eclipse.

The measurements of the fracture toughness (K_{IC}) were found to be 2.31 MPa√m for Meliodent and 3.26 MPa√m for Eclipse resins.

Results obtained by the strength indentation (IS) method are comparable to those obtained by the SENB method at low loading rates (~0.05 mm/min), as shown in **Table 3**.

3. Studies concerning biocompatibility of acrylic resins

Acrylates are well known for their potential allergies. To evaluate the allergy potential of acrylic resins, we used *in vivo* and *in vitro* tests. For *in vivo* testing, we have used the patch test and the tumour necrosis alpha-factor (TNF) test, which is eloquent for cytotoxic detection and can detect early signs of allergy and inflammation by measuring endotoxins in patient serum. Acrylate toxicity is mainly due to the residual monomer. Therefore, we have also carried out tests to determine the amount of residual monomer present in the resin by the volatile-component-content method and bromine index methods [11, 12].

3.1. *In vivo* testing

3.1.1. *Patch test*

Patch testing involves applying patches dipped in supposedly antigenic substance on a free-rash segment and off the rash periods and maintaining them in contact with the skin for 48 h. The area for testing is the mid-upper back, anterior sides of forearms and upper external region of the arms, as shown in **Figure 14** [13].

IQ Ultra Chambers Box (Chemotechnique Diagnostics) tests were used, with methyl methacrylate as an allergen. All 10 subjects showed tenderness (sensibility) to methyl methacrylate [14].

3.1.2. *TNF alpha test*

TNF alpha (tumour necrosis alpha-factor) test which is eloquent for cytotoxicity detection was used in three cases of mucosal reactions in patients wearing acrylic dentures. The enzyme-linked immunosorbent assay (ELIZA) method was applied to the patient's serum [15]. Five hundred microliters were harvested from each patient (having a duplicate sample) in order to determine the TNF alpha concentration. Quantikine Human TNF alpha immunoassay which

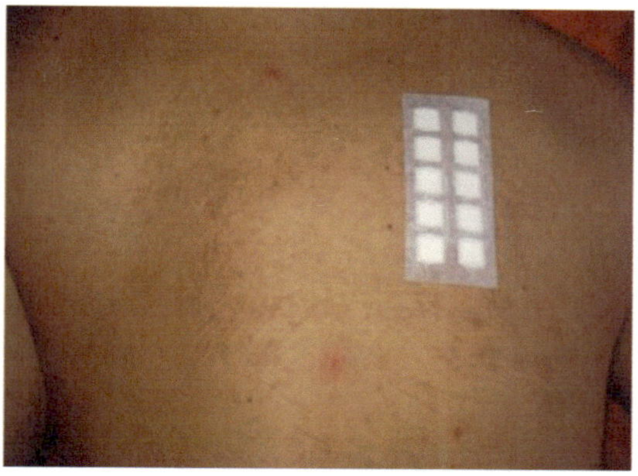

Figure 14. Patch test.

allows the quantitatively determination of cytokines on patient serum was used. The substrate solution was obtained by mixing equal volumes of A and B reagents from the kit, maximum 15 min before usage. The supply solution (the undiluted standard) was prepared by reconstructing the cytokine standard using calibrator thinner RD6-21. For standard preparation, 500 μL of calibrator thinner RD6-21 was used. After pipetting 500 μL of solution in the first tube, the mixing concentration was calculated. The solution was well stirred and 500 μL was transferred into the next tube. The concentration is once again calculated. Standards for different concentrations are prepared in the same way (500, 125, 62.5, 31.2, 15.6 pg/mL). The supply solution was used as high standard (1000 pg/mL) and the calibrating thinner RD6-21 was used as zero standard (0 pg/mL). Note that 100 μL of thinner RD1-51 was pipetted in each hole of the plate and 100 μL of each standard was pipetted in the first strip in the following order: 0, 15.6, 31.2, 62.5, 125, 250, 500 and 1000 pg/mL. Starting with the second strip, 100 μL was pipetted. The plate was covered with an adhesive film and incubated for 2 h at room temperature. After that, the plate was washed automatically four times with 400 μL of washing solution and placed on an absorbent paper to clean the samples without damage. Note that 200 μL of cytokine conjugate was added in each hole and the plate was covered with an adhesive film and incubated for 2 h at room temperature. After the incubation period, the adhesive film was removed and the plate washed in the same way. Note that 200 μL of substrate solution was then added to each hole of the plate, the plate was covered again with an adhesive film and incubated for 30 min at room temperature, in darkness. After incubation time was over, 50 μL of stopping solution was added in each hole of the plate. When the reaction stops, the colour turns from blue to yellow. If the colour is green or does not modify uniformly, the plate will be slightly stirred for complete homogenization. The optical densities are determined 30 min after the reaction was stopped using an automatic reader at a wavelength of 450 nm, with a reference filter of 540 or 570 nm (for correction) (**Figure 15**). In the case of samples that are read only at 450 nm without correction, the accuracy may be influenced. Results are obtained depending on the logarithmic calibration curve of each cytokine, build on absorption and standard concentration (0–1000 pg/mL). The results are shown as optical density units auto-

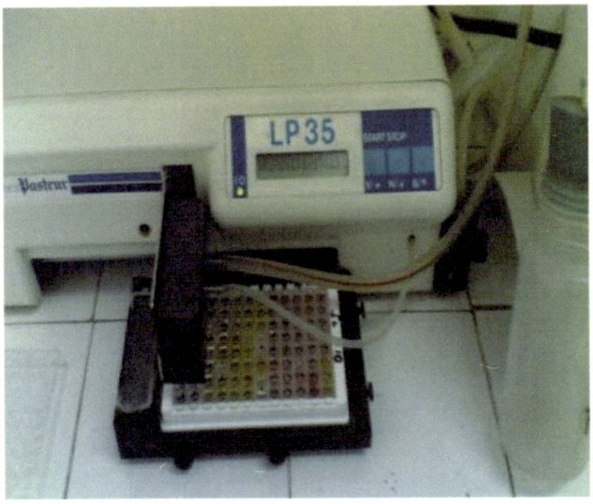

Figure 15. Results reading for TNF alpha test.

matically converted into pg/mL for each of the sample tested. Determination of TNF-alpha concentration in the first case revealed slightly increased values of this cytokine (50.48–50.7 pg/mL). In the second case, the values were higher (90.3–90.9 pg/mL) and in the third case the values were the highest (100.4–107 pg/mL), which indicate certain inflammation. The results obtained with TNF-alpha test are eloquent for cytotoxic detection of early signs of allergy and inflammation by measuring the endotoxins from the patient serum. It is always recommended to associate multiple types of clinical tests and these should always be histologically confirmed [16].

3.2. *In vitro* testing

3.2.1. Volatile-component content method

Three acrylic resin samples were used, each one being harvested from a different full denture base. The samples were preconditioned at 80°C for 2 h for removing moist. The weight of the samples is S1: 0.7863, S2: 0.05638 and S3: 0.8421 g. A Petri box is sterilized at 150°C for 1 h and then weighed in the analytical scale with high precision. The samples are positioned in the Petri box and weighed again very accurately. They are kept in the oven for 10 h at 150°C, and weighed again, very precisely. The weight difference is given by the amount of existing residual monomer: S1: 0.7864 g, S2: 0.5640 g, S3: 0.8422 g. In all these three cases, no significant weight loss was noticed. These results conclude that almost no residual monomer was found in the three full dentures [14].

3.2.2. Bromine index method

In order to verify the above results, we have used the bromine index method which determines the percentage of monomer in the sample, based on the amount of bromine added to the double bond (C=C) links. All the three samples were found to be free from the residual

http://dx.doi.org/10.5772/intechopen.69010

monomer. We have concluded that in the three PMMA samples, there is no residual monomer in the acrylic resin samples. But, there is less than a 0.0000 g order which is not detectable. This shows that a very accurate manufacturing method was adopted in the case of the three considered full acrylic denture bases [17].

4. Alternative resins and technologies for denture manufacturing

Our experience in denture manufacturing includes a variety of alternative resins and manufacturing technologies such as self-curing acrylics manufactured by casting, different thermoplastic resins manufactured by injection, light-cured diacrylic and urethane-based resins and poly(ether ether ketone) (PEEK) high-performance polymers manufactured by milling.

4.1. Self-curing acrylics manufactured by casting

Full-denture casting represents one choice to classic heat-curing acrylic dentures. Self-curing acrylic resins suitable for casting belong to type 2, group 2 of acrylic resins (**Table 1**). For polymerization, temperatures below 65°C are used. The acrylic paste was previously prepared in a texture suitable for casting and then poured in a special flask. The mould is made up of either reversible hydrocolloid or silicone (**Figure 16**), compared to the classic plaster mould used for investing heat-cured acrylic dentures. The casting system has the following advantages: the reversible hydrocolloid can be reused and the polymerization time is shorter. There are wide colour ranges (10 colours) of the acrylics and minimal adjustments are required. The most common errors when using this technology are consequence of bubbles forming when pouring the mould. This causes porosity of the mucosal surface and lack of substance due to the fast setting of the resin. Therefore, great skill is required when pouring the mould. Porosity may also occur due to improper preparing of the acrylate [4, 18].

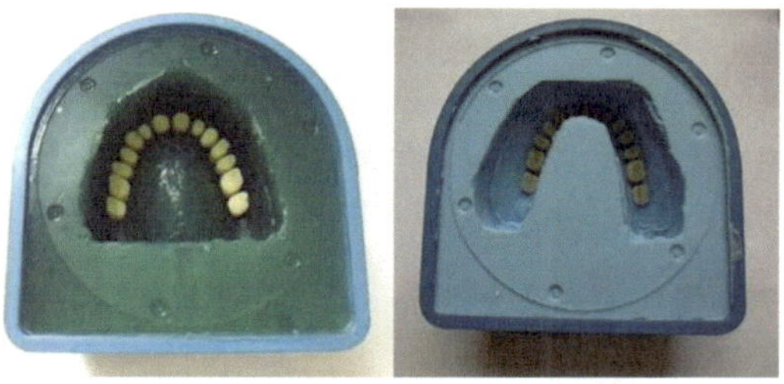

(a) (b)

Figure 16. The mould made up of (a) reversible hydrocolloids and (b) silicone, before pouring the acrylic paste.

4.2. Thermoplastic resins manufactured by injection

Compared to classical self- or heat-curing acrylates, thermoplastic resins have a number of advantages as follows: a very good long-term performance, maintaining their size and colour in time, stability, resistance to deformation, wear and solvents, very good tolerance due to the absence or reduced quantity of residual monomer, responsible for allergies in a lot of patients, no porosity which prevents development of microorganisms and deposits [4].

The advantages of the injecting system lie in the fact that the resin is delivered in a cartridge (**Figure 17a**) which eliminates dosage errors, guaranteeing long-term stability of the shape, reduced contraction, as well as mechanical resistance with ageing. The disadvantages are mainly the consequence of the high cost of the injection device (**Figure 17b**) and of the materials to be used.

The processing technology implies the thermal plasticization of the material, in the absence of any chemical reaction and injection of the plasticized resins into a mould [3].

Advantages of thermoplastic resins include removable partial dentures, preformed clasps, partial denture frameworks, temporary prosthetic restorations, full dentures, orthodontic appliances, anti-snoring devices, mouth guards and splints.

Thermoplastic resins include acetal, polycarbonate (polyesters group), acrylates and polyamides (nylons) [19].

Metal-free removable partial dentures made up of thermoplastic materials represent a modern alternative solution to the classical metal framework dentures. These have the advantages of being lightweight, flexible and much more comfortable to the patient. These are also biocompatible, non-irritant, sure, non-toxic, biologically inert, superior aesthetics and also offer quality static and dynamic stability [5]. The clasps are made up of the same material as the denture base or are readily-made from the same material. Though the mechanical resistance is most important, the first choice for manufacturing the framework is an acetal resin. The removable partial dentures with acetal resin framework are the most laborious to manufacture. The acetal framework was being manufactured first, followed by the acrylic saddles and artificial teeth (**Figure 18**).

These types of partial dentures have thin frameworks, with flexible and aesthetic clasps [20].

Thermoplastic polyamide (nylon) is a versatile material, with high flexibility, physical strength, heat and chemical resistance. The super flexible polyamide is extremely elastic, virtually unbreakable, lightweight and impervious to oral fluids. The medium-low flexibility polyamide is a half-soft material which offers superior comfort, good aesthetics and could be

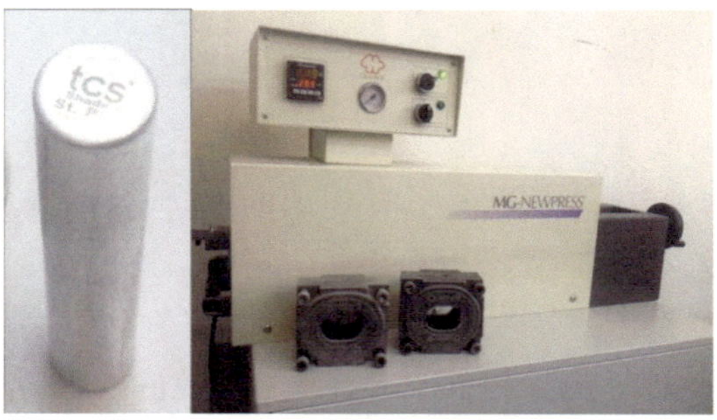

(a) (b)

Figure 17. (a) The cartridge of thermoplastic material and (b) the injection unit MG-Newpress (Quattroty).

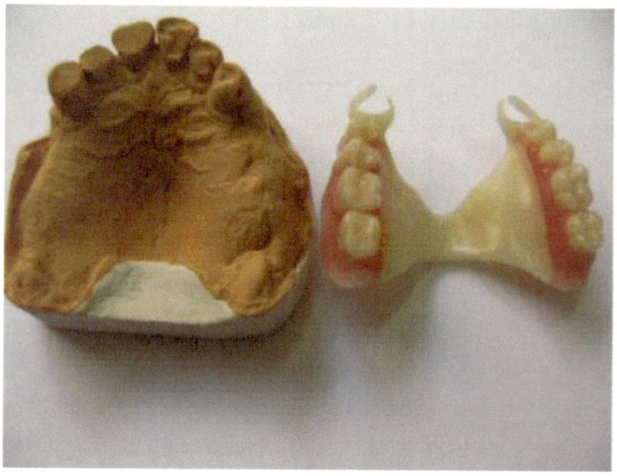

Figure 18. Partial denture with acetal framework, clasps and acrylic saddles.

used for removable partial dentures. Super flexible polyamide is especially useful for retentive dental fields, which would normally create problems with the insertion and disinsertion of the removable partial dentures. The clasps are made up of the same material as the denture base in the case of super flexible polyamide dentures. In the case of medium-low flexibility polyamide dentures, ready-made clasps are an option. Metal clasps may also be used [21] (**Figure 19**).

Polycarbonate resins are particularly polyester materials. These have good fracture strength and flexibility, a natural translucency, but the wear resistance is lower than acetal resins and are not recommended for partial denture frameworks. The finishing is very good, which makes them suitable for temporary prosthetic restorations.

Thermoplastic acrylate has the highest impact rating of any acrylic, has long-term stability, its surface structure being dense and smooth. This material was developed for manufacturing complete dentures. It is not elastic, but its flexibility makes it practically unbreakable as one can bounce such a denture off the floor without cracking the base. The biocompatibility is very good due to the absence of residual monomer. The denture has very good long-term stability because water retention is limited.

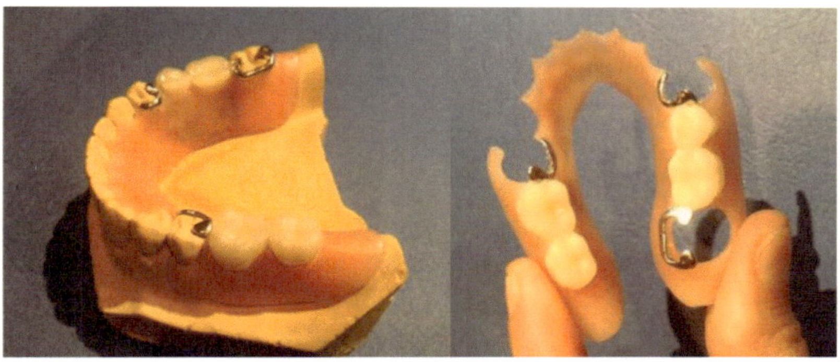

Figure 19. Removable partial denture made up of superflexible polyamide with metal clasps.

4.3. Light-cured diacrylic and urethane-based resins

Diacrylic composite resins are complex materials. Initially elaborated as aesthetic restorative materials, these developed a lot. The advantages of prosthetics are as follows: veneering of metal-polymeric fixed dentures, single-tooth or temporary crowns, inlays, onlays, epitheses, repairing damaged porcelain veneers, artificial teeth, base of removable dentures and repairing removable dentures. Compared to acrylics, these have a lower shrinkage during poly-merization and have superior physico-mechanical and chemical resistance. Diacrylic composite resins can be made using self-, heat or light-curing, in some cases several curing methods being combined. Light-curing diacrylic resins are successfully used in dental laboratories, especially for veneering, having the advantage of prolonged handling time. In addition, these physico-chemically adheres to the metallic framework, have good colour stability in time and a special aesthetic effect, in part due to the wide selection of shades available for veneering. These can be easily repaired after fixing in the oral cavity, if needed [1].

The absence of methyl, ethyl, propyl and butyl groups in urethane-based resin composition does not generate contact allergies. The light-curing Eclipse Resin System (Dentsply-DeguDent) allows a rapid manufacturing of full dentures, eliminating some time-consuming intermediate steps, like investing and heat-curing. The light-curing resins of Eclipse Resin System contain aliphatic urethane dimethacrylate-urethane oligomers (UDMA) as base monomers and acrylic copolymers, an inorganic submicronic silica filling, a light-curing initiating system and additives. The system consists of three types of resins, which can be handled like wax (base plate, set-up and contour resins). The light-curing protocol was made available by the producer in many variants, which correspond to different technical procedures. The system is extremely efficient, a complete denture base may be ready in 30 min, after master model complete setting. The 'wax-up' is practically made on the denture's polymerized base and, after checking it, the rest of the pattern (saddles) was light-cured. Thereafter, the denture was finished [6] (**Figure 20**).

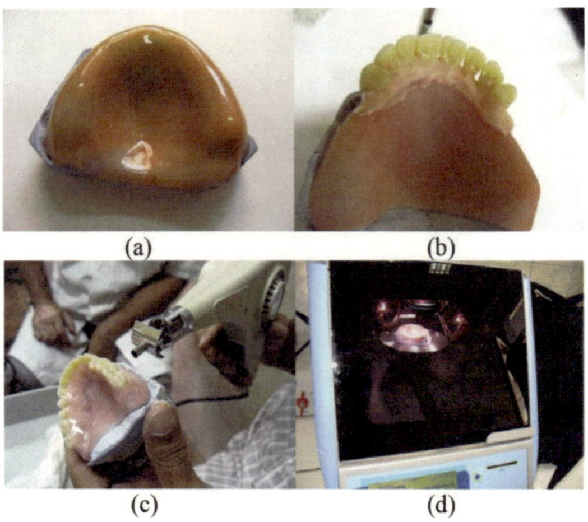

Figure 20. (a) Base plate resin before light-curing, (b) teeth mounting using set-up resin, (c) contour resin processed using the warm air gun and (d) light-curing unit in use.

http://dx.doi.org/10.5772/intechopen.69010

4.4. High-performance polymers, manufactured by milling

Poly(ether ether ketone) (PEEK) is a high performance polymer used in dentistry since 2002. PEEK is a biocompatible thermoplastic material, with superior properties such as mechanical properties, resistance to wear and fracture and elasticity comparable to bone. PEEK for dental use may be optimized by adding ceramic 0.3–0.5 μm particles. It is resistant to high temperature, good stability, easy to be polished properly, insoluble in water and ideal for allergic patients. The material (grains) may be injected at 400°C or milled (disks) using a CAD/CAM system (**Figure 21a**). It is indicated for crowns and bridges (replacing the metal framework), abutments and removable partial dentures including precision attachments [22, 23].

A removable partial denture framework made up of Juvora PEEK (Invibio) (including clasps), using the Exocad CAD and Coritec 450i (imes-icore) CAM, weights only 1.36 g (**Figures 21b** and **22a**). The saddles are made of acrylate having the entire denture weights only 3.36 g (**Figure 22b**).

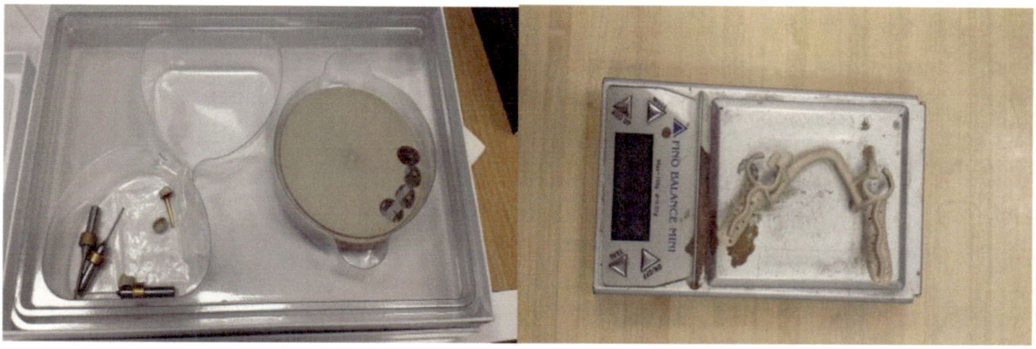

(a) (b)

Figure 21. (a) PEEK milling disk and (b) PEEK framework on a weighting scale.

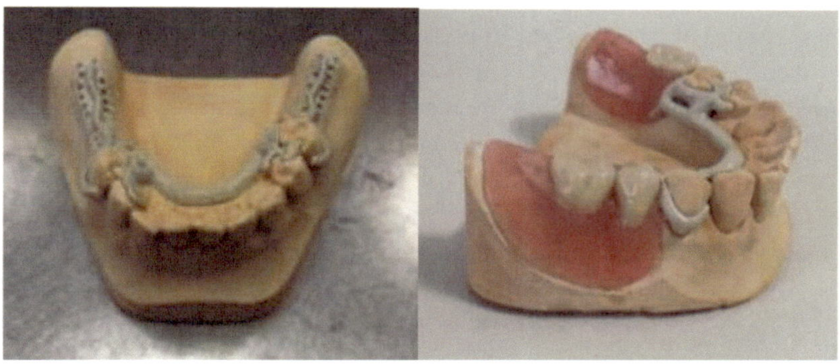

(a) (b)

Figure 22. (a) PEEK framework and (b) finished denture.

5. Conclusions

New choices of resins, with better properties compared to acrylics, are now suitable for dental applications. Alternative technologies for processing dental resins, like casting, injection, light-curing and milling are meant to improve their clinical performances.

Long-term deterioration of resin-based dentures in oral environment is still an unsolved problem. This can be improved by bringing innovations in the manufacturing technology, thereby reducing polymer defects and distortions in a warm and humid environment, material's fatigue and ageing. Moreover, we have to consider various possible environmental aggressions such as acids, alcohol, tobacco and thermal fluctuations, the continuous contact with water and salivary enzymes, food remnants, fluctuation of the oral balancing system, the contact with live tissues, bacteria and yeasts. It seems that the enzymes from human saliva are able to produce the softening of the resin surface, probably by hydrolyzing the small oligomers into monomer phases [8].

Choosing the right material for manufacturing partial or full dentures is very important because it has direct effect on their quality and lifetime. Because they are brittle, dentures made up of acrylic resins have a limited life in the mouth. Moreover, following some manufacturing procedures, small defects (holes in the polymer structure) may occur. The continuous mastication stress (repeated movements of low amplitude) and the oral environment have an important role in the degradation of the denture in time [9].

The fracture toughness depends on the type and the nature of the polymer and the reinforcement added components. For example, fracture toughness of a monomethacrylate-based material is lower than that of dimethacrylate-based material. An increase in the fracture toughness can be achieved by adding reinforcement fibres which prevent or slow down the crack growth or by adding rubber-like substances, which increase elasticity [9].

Currently, the selection of dental materials is made particularly in terms of manufacturing technology and aesthetic criteria. These criteria are not sufficient to provide functional, durable and comfortable dentures. In the case of acrylic resins used in dental prosthetics, in addition to aesthetic aspect and strength, elasticity and elongation are also important features. Biomechanical studies of dental resins, implying different technologies, are necessary in order to discover potential causes of failure. These are not only the consequence of material defects and solubility, but can also depend on the technological procedures and processing. Most of the resins distort and fracture at low level of tension dependent on the environmental or loading conditions. In our studies, the values of assessed mechanical properties were lower for the wet samples than for the dry samples, indicating a clear influence of saliva in the biodegradation of the material, with direct consequences biomechanical performances and lifetime for dentures [24, 25].

As far as the allergy potential of acrylic resins is concerned, our results show as follows: *in vivo* tests, both patch test and TNF-alpha showed allergic reactions to acrylate. Both *in vitro* determinations for residual monomer showed a very small and in fact practically undetectable amount in our samples. This led to the conclusion that the manufacturing process was carried out with maximum thoroughness.

Resin-based dental materials have a wide range of applications. But, despite the efforts made to continuously improve their physical, mechanical and aesthetic properties thereby cause side effects regarding their biocompatibility. These situations may lead to local lesions which can prove extremely unpleasant for the patient [16].

In our opinion, the best results in denture manufacturing may be achieved by combining scientific principles with creativity, while the optimal choice of the material and technique has a major importance.

Author details

Lavinia Cosmina Ardelean[1]*, Cristina Maria Bortun[2], Angela Codruta Podariu[3] and Laura Cristina Rusu[1]

*Address all correspondence to: lavinia_ardelean@umft.ro

1 Department of Technology of Materials and Devices in Dental Medicine, "Victor Babes" University of Medicine and Pharmacy, Timisoara, Romania

2 Department of Dentures Technology, "Victor Babes" University of Medicine and Pharmacy, Timisoara, Romania

3 Department for Preventive Dentistry, Community Dentistry and Oral Health, "Victor Babes" University of Medicine and Pharmacy, Timisoara, Romania

References

[1] Ardelean L, Bortun C, Motoc M, Rusu L, Motoc A. Errors in full denture casting using acrylic resins. Revista Materiale Plastice. 2008;**45**(2):214-216

[2] Ardelean L, Bortun C, Motoc M, Rusu LC. Alternative technologies for dentures manufacturing using different types of resins. Revista Materiale Plastice. 2010;**47**(4):433-435

[3] Ardelean L, Bortun C, Motoc M. Metal-free removable partial dentures made of a thermoplastic acetal resin and two polyamide resins. Revista Materiale Plastice. 2007;**44**(4):345-348

[4] Ardelean L, Bortun C, Podariu A, Rusu L. Manufacture of different types of thermoplastic. In: El-Sonbati AZ, editor. Thermoplastic Composite Materials. Rjeka: InTech; 2012. pp. 25-48

[5] Ardelean L, Bortun C, Podariu AC, Rusu LC. Some alternatives for classic thermopolymerisable acrylic dentures. Revista Materiale Plastice. 2012;**49**(1):30-33

[6] Ardelean L, Bortun CM, Podariu AC, Rusu LC. Thermoplastic resins used in dentistry. In: Das CK, editor. Thermoplastic Elastomers. Synthesis and Applications. Rjeka: InTech; 2015. pp. 145-167. DOI: 10.5772/59647

[7] Ardelean L, Reclaru L, Bortun CM, Rusu LC. Assessment of dental alloys by different methods. In: Aliofkhazraei M, editor. Superalloys. Rjeka: InTech; 2015. pp. 141-170. DOI: 10.5772/59358

[8] Ardelean L, Rusu LC, Bratu DC, Bortun CM. Diacrylic composite resins as veneering materials. Revista Materiale Plastice. 2013;**50**(2):93-96

[9] Biclesanu C. Dental pain. Revista Romana de Stomatologie. 2012;**58**(2):77-78

[10] Bolos OC, Bortun CM, Cernescu A, Ardelean L, Bolos A, Rusu LC. Fracture toughness evaluation of some resins used in complete dentures technology. Revista Materiale Plastice. 2013;**50**(1):28-31

[11] Bortun C, Cernescu A, Ghiban N, Faur N, Ghiban B, Gombos O, Podariu AC. Durability evaluation of complete dentures realized with "Eclipse Prosthetic Resin System". Revista Materiale Plastice. 2010;**47**(4):457-460

[12] Bortun C, Ghiban B, Sandu L, Faur N, Ghiban N, Cernescu A. Structural investigations on mechanical behavior of two dental acrylic resins. Revista Materiale Plastice. 2008;**45**(4):362-366

[13] Bortun CM, Ardelean L, Rusu LC, Marcauteanu C. Importance of modern light-curing resins in the design of removable partial dentures. Revista de Chimie. 2012;**63**(4):428-431

[14] Bortun CM, Cernescu A, Ardelean L. Mechanical properties of some dental resins in wet and dry conditions. Revista Materiale Plastice. 2012;**49**(1):5-8

[15] Ciceoi AI. Removable partial denture with non-metallic framework [thesis]. Timisoara: The Victor Babeş University of Medicine and Pharmacy; 2016.

[16] Faur N, Bortun C, Marsavina L, Cernescu A, Gombos O. Durability studies for complete dentures. Key Engineering Materials. 2010;**417-418**:725-728

[17] Ghiban N, Bortun CM, Bordeasu I, Ghiban B, Faur N, Cernescu A, Hanganu SC. Evaluation of mechanical properties by stereo-and scanning electron microscopy of some heat curing dental resins. Revista Materiale Plastice. 2010;**47**(2):240-243

[18] Mermeze AI. Rehabilitation of partial edentation with a flexible denture with metallic clasps. [thesis]. Timisoara: The Victor Babeş University of Medicine and Pharmacy; 2016.

[19] PEEK. A New Material for CADCAM Dentistry [Internet]. 2014. Available from: https://juvoradental.com/en/2014/0613/peek-a-new-material-for-cadcam-dentistry [Accessed: 24 February 2017]

[20] Podariu AC, Ardelean L, Jumanca D, Galuscan A, Rusu LC. Determining the amount of volatile organic phase in PMMA dentures. Revista de Chimie. 2012;**63**(7):720-721

[21] Podariu AC, Jumanca D, Galuscan A, Podariu AS. Determination of fluor cytotoxicity in combination with cholecalciferol. Revista de Chimie. 2012;**63**(12):1249-1250

[22] Reclaru L, Ardelean L, Rusu L. Toxic materials, allergens and mutagens and their impact on the dental field. Medicine in Evolution. 2008;**14**(3):98-102

[23] Rusu LC, Ardelean L, Podariu AC, Matei C, Tampa M. Allergenic potential evaluation of acrylic resins from the complete prostheses. Revista Materiale Plastice. 2012;**49**(2):133-134

[24] Rusu LC, Ardelean L. CAD/CAM technology concerning biocompatibility in zirconia all-ceramic restorations. Revista de Chimie. 2012;**63**(5):513-515

[25] Rusu LC, Urechescu H, Ardelean L, Levai MC, Pricop M. Comparative study for oral reaction produced by polymethylmethacrylate. Revista Materiale Plastice. 2015;**52**(3):413-415

Chapter 2

Acrylates in Dental Applications

Bożena Tyliszczak, Anna Drabczyk and
Sonia Kudłacik-Kramarczyk

Additional information is available at the end of the chapter

http://dx.doi.org/10.5772/intechopen.69008

Abstract

In the presented chapter, the role that is played by acrylates in dentistry has been characterized. In the introduction, subject of oral diseases has been raised as well as an issue of development of dentistry over the centuries. Furthermore, characteristics of the materials that have been used over the years to receive elements used in the form of prosthetic devices or dental implants that in the most favourable way from the user's point of view enable the restoration of the missing piece of the dentition have been performed. Next, composition, functions and types of teeth have been described. In the following sections, materials (including dentures, adhesives, impression trays and dental crowns) widely used in dentistry and dental prosthetics in the preparation of which the key role is played by acrylates have been characterized. The preparation of prostheses was described. Particular attention has been drawn on the possibility of modification of the synthesis of acrylic materials that can lead to the improvement of their properties and result in making them more favourable from the point of view of the patient. The chapter is crowned with a brief description of the studies of properties, which are subjected to dental materials before application in the dental office.

Keywords: acrylates, dentistry, dentures, dental fillings, strength tests of teeth

1. Introduction

Tooth decay, which is undoubtedly associated with dental sciences, was the cause of a great progress and development in that field. It is a disease that causes permanent destruction of the tissues of the tooth. Therefore, a lot of research was conducted in order to create materials appropriate for restoration of those losses. Particularly in the last quarter, there was a very great progress in the field of dental materials science [1–3].

Complications related to the tooth decay can lead to many diseases of pulp and periodontal. Therefore, scientists are still carrying out research on new materials, therapeutic techniques and medicines. Currently, on the market of medical supplies, it is possible to distinguish materials that:

- Eliminate and prevent pathological conditions.
- Prevent chemotoxic iatrogenic states.
- Treat disease states.
- Stimulate regeneration processes [1–3].

The problem of reconstruction of lost parts of the body as a result of an accident or an illness accompanies human life since time immemorial. The same problem applies to the dentistry. Replacement of tooth structure constitutes an extensive section of dentistry and is necessary from an aesthetic and functional point of view. Some important factors have an impact on the process of reconstruction as well as on achieving a satisfactory effect. Such factors should be mentioned as the availability of suitable materials, growth of technology and procedures used during application and development of fields of science such as chemistry, biology and physics, without which the progress of dentistry would be impossible. In order to enable a better understanding of the above-mentioned considerations in **Figure 1**, dependencies of various sciences applied in dentistry are presented.

The principles of physics, chemistry and biology are certainly used when comparing the physical and structural characteristics of restorative materials as well as during their application. The knowledge of these principles is very important for every dentist because with their help, it is possible to understand the basic phenomena such as melting or cooling of alloys or formation of crystalline structures in tested materials. It is also important to be able to predict

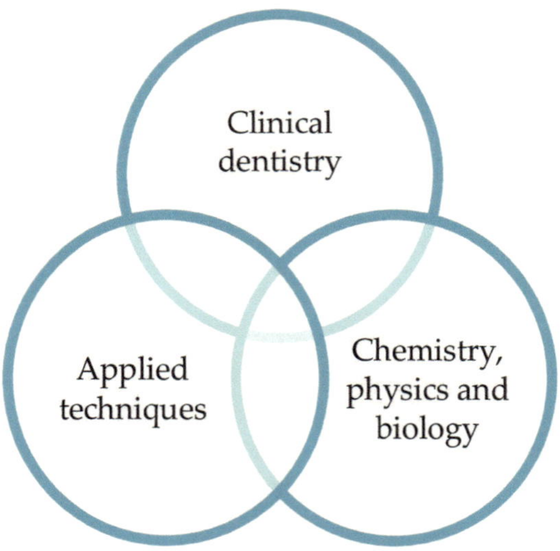

Figure 1. Schematic relationship between the sciences and the development of dentistry.

the chemical reactions which can occur during reconstruction of defects by using different materials or alloys. Of course from the clinical point of view, it is essential that conducted research was as practical as possible. However, practical clinical dentistry and scientific theory will not demonstrate desirable progress and will not become mutually friendly if these areas will not be developing in parallel drawing on mutual reports [4–6].

Continuing the discussion on the development of dentistry, an important issue is undoubtedly a history of this field of medicine, which began several hundred years before the period of Christianity. The first actions in the field of dentistry include the receipt of dentures on the basis of gold by Phoenicians. Subsequently, metals such as tin, copper, silver and lead began to play an important role in the synthesis of dental materials. In 990 B.C., methods of iron processing have been identified. Currently, scientists do not have any information about who and in what way constructed prehistoric dental restorations prevailed at that period. Certainly we can say that the role played during that period by goldsmiths and artisans was as important as that one plays nowadays by dental technicians [4–6].

As for the practice of using dental bridges and crowns, it is likely to be developed in ancient Rome around 500 B.C. The ability of carrying out such treatments is associated with the fact that people possessed appropriate skills of soldering and riveting at such a level that conducted operation did not cause bodily injury. It is obvious that craftsmen and other persons performing such treatments must be characterized by thorough understanding of melting process and execution of fluxes [4–6].

Further development of civilization apart from dental restorations and application of the aforementioned dental bridges and crowns attempts in which improperly functioning tooth was replaced by tooth of another patient or by tooth derived from the animal have been taken. It was based on immobilization of the tooth in the mouth by means of a wire. It should also be noted that ancients personally produced the mentioned wire [7].

Period from 1600 to 1840 was considered as the beginning of dental science. Previously, this field of medicine was perceived as a craft made mostly by craftsmen. In 1746 on a market, a book of Claude Munton entitled *Essay on the dental techniques or dissertation on artificial teeth* appeared and treated with a mechanical aspect of dentistry. In that publication, gold-based buckles and dental crowns that maintained the artificial teeth in the mouth have been described. During the following years, a number of texts about dentistry have been written in which it was mentioned about teeth of ivory, gold pins used for maintaining teeth as well as about porcelain as a material for the preparation of teeth and platinum that was used for the synthesis of dental hooks [7, 8].

Work of R.C. Skinner that was written in 1801 is considered as a first textbook about dentistry. In that time, dentistry began to be perceived as a professional field of science created by dentists and surgeons. In subsequent years, the progress and development of other sciences were accompanied by an invention of the amalgam, which was considered as one of the greatest and most important discoveries in the field of restorative materials (**Figure 2**). Furthermore, a lot of dissertations on the dental subjects have been created, as well as new technologies have been developed at a satisfactory pace [9].

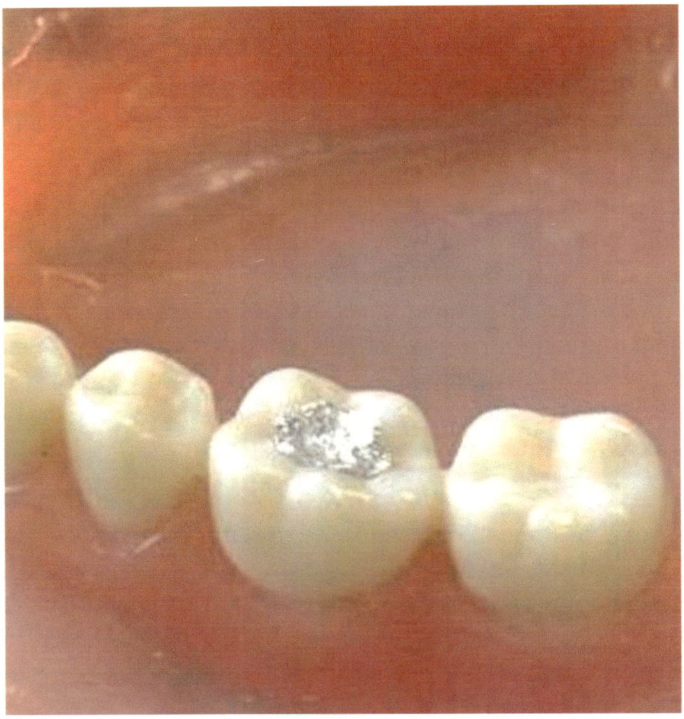

Figure 2. Amalgam as a dental filling.

In the second half of the nineteenth century, a large development of the field of plastic was observed. In 1869, a new material that was called 'celluloid' was presented. In 1871 that material was used as a restoration which was a turning point for the introduction of plastics for the dental industry. Despite its advantages such as ease of handling or resistance to fracture, new material was also characterized by many disadvantages including an undesirable colour change and emission of mercury causing drooling and inflammations in the mouth. Thus, in conjunction with further explorations of new material, next materials that have been proposed by researchers were phenol-formaldehyde resins [10]. As in the case of celluloid despite initial delight of dentists in these resins, it became quickly apparent that they do not meet the basic requirements. Thus in 1935, ICI company unveiled a new material which was poly(methyl methacrylate) (PMMA) also known as 'organic glass' which is associated with its transparent properties. At first, PMMA was not applied in dentistry due to many technological problems. However, in 1937, a new method of synthesis of the mentioned polymer was developed in which PMMA was obtained in the form of beads and not in the form of plates as it was previously the case. Then, a special methodology was developed aimed at combining beads of polymer with monomer in order to obtain an elastic mass. In **Figure 3**, PMMA in a form of plates (A) and beads (B) is presented.

Detailed application of this polymer and the other belonging to the group of acrylates will be presented in subsequent parts of this chapter.

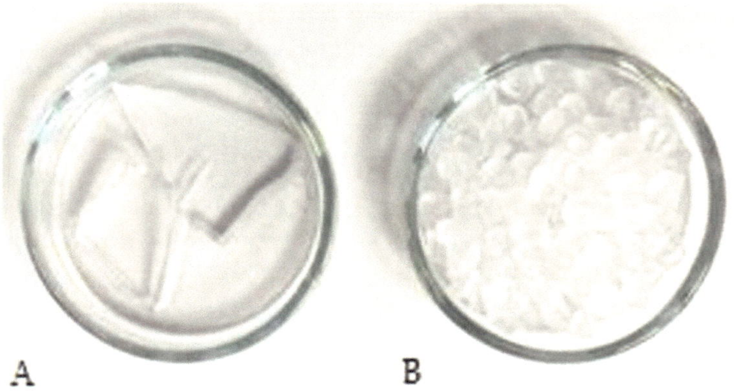

Figure 3. PMMA in a form of plates (A) and beads (B).

2. Anatomy of the tooth

The name 'teeth' refers to the anatomical formations of complex structure occurring in the oral cavity that play an essential role in the digestive system. Teeth fulfil many functions including grinding and crushing of food which significantly facilitates its subsequent digestion and intake in the following sections of the digestive tract [11] (**Figure 4**).

Figure 4. Acrylic teeth.

In human oral cavities, it is possible to distinguish two types of dentitions: primary and permanent one. First of the mentioned occurs in children aged 2–6 and involves incisors, canines and molars. Primary dentition is with the time replaced by permanent one. This type includes 32 teeth, and in comparison to the previous one, it comprises additionally premolars [11]. The different types of teeth in dentition were shown in **Figure 5**.

In the construction of every tooth, three basic elements can be mentioned, i.e. crown, neck and root. Among tissues that form teeth, enamel, dentin, cementum, and pulp have to be mentioned. Enamel—the toughest tissue in the human body—in the vast majority (95%) is made of inorganic compounds. The most innervated and vascularized tissue from the above-mentioned is a tooth pulp [11, 12].

Inadequate oral hygiene which consequently results in the accumulation of food debris and in the development of bacterial flora leads to the development of various types of teeth and periodontal diseases. In some cases, the process of healing of this type of illnesses is complex and lengthy. Therefore, it is necessary to develop this field of medicine whose role is to deal with functioning but also with treatment of teeth, periodontal, all tissues and any other elements contained in the oral cavity. The origins of this science should be sought several thousand years ago. Literature reports indicate that the traces of actions aimed at preventing tooth decay were discovered in the skull of young man whose remains were found more than 14,000 years ago. Areas of dentistry such as orthodontics and prosthetics also are characterized by a long tradition. The hygiene and the aesthetics of oral care were taken in ancient times. Therefore, high attention has been paid to the development of this field of medicine. With the passing of years, new elements that had to meet the growing demand on the market of dental materials have been introduced [11, 12].

Nowadays, it is necessary to use dental materials associated with artificial restorations or with technologies that in perfect way will enable a reconstruction of the missing piece of the dentition. Such materials are undoubtedly dental bridges, crowns, dentures and dental braces [11, 12].

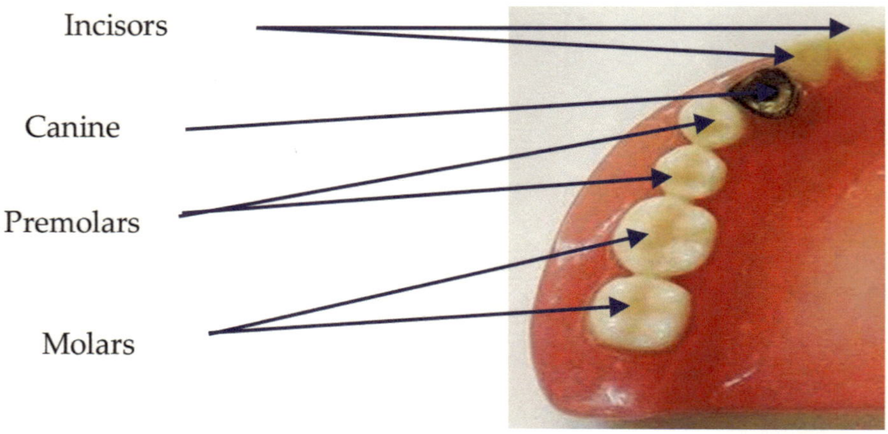

Figure 5. Kinds of teeth in permanent dentition.

3. Bonding materials in dentistry (cements and fillings)

The concept of 'cement' is associated unambiguously with the construction industry. However, it should be noted that this term also refers to the materials constituting the binder in dentistry. There is a wide diversity of dental cements including zinc-phosphate, glass-ionomer, compomer or composites (on the basis of resin). Last of the mentioned due to their characteristics are becoming more and more popular on the market of medical products [13].

Composites based on acrylic resins are materials widely used for dental restorations. These are becoming increasingly popular displacing from the market of medical products amalgam fillings. The use of those fillings caused the problem in terms of safety due to the fact that their main component is toxic mercury. However, it should be noticed that in some ways, they outweigh the fillings on the basis of acrylate resins. In fact, they were characterized by greater durability and better resistance to fracture. Therefore, nowadays, researchers are working on the improvement of the physico-chemical properties of resin based on acrylates [13].

Acrylic materials are included in glass-ionomer cements. This filling is composed of two elements, i.e. powder and liquid. The liquid is an aqueous solution of copolymers of itaconic and polyacrylic acids, while powder contains fluoro-calcium-aluminium-silicon glass. Interesting feature that distinguishes this type of cement from the others on the market of dental products is a very good adhesion to enamel and dentin. Furthermore, glass-ionomer cements are characterized by an ability of forming a chemical bond with the tissue of the tooth which contributes to the cariostatic action of such filling (protection against tooth decay). Polyacrylates contained in the liquid react with the compounds forming the powder and consequently release of various types of ions with a predominance of fluoride anions occurs. As a result of their reaction with polyacids, formation of a gel matrix occurs which combines chemically with the dentin. This is a middle layer of a tooth which is composed of fluid-filled tubules, connective tissues and nerves. The elements included in the dentin are mainly calcium (27%) and phosphorus (13%) [14].

Combining of glass-ionomer cement (**Figure 6**) with the tooth is a result of forming a chemical bond between ions contained in cement and ions of phosphorus and calcium included in the structure of enamel or dentin. Tooth surface must be previously cleaned that has an impact on the effectiveness of the formed combination [14].

Glass-ionomer cement is characterized by good resistance to bending. However, these materials have also some disadvantages including low tensile strength. Undoubtedly, one of the most favourable features of these fillings is their antibacterial activity which derives from the presence of fluorine in their structure [15].

Hybrid cements such as glass-ionomer fillings or those one modified with resin are characterized by slightly better strength properties. It is worth noting that these materials are chemo or light cured. Besides liquid and powder, such cements contain in their composition catalysing system of a micrometric size disposed in the form of capsules. The mentioned system consists of ascorbic acid and potassium persulfate, whereas liquid comprises an aqueous solution of polyacrylic acid containing additionally 2-hydroxyethyl methacrylate (HEMA). The connection

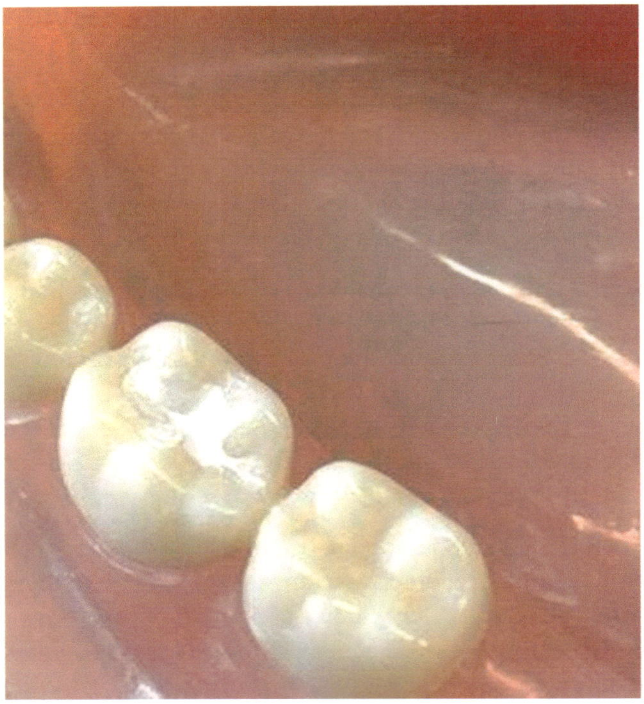

Figure 6. Glass-ionomer cement.

between this kind of filling and a tooth is stronger that a bond that is formed by glass-ionomer cement [16, 17].

Acrylates play an important role also in the compositions of zinc oxide-polyacrylic cements (known also as zinc oxide-polycarboxylate cements). The aqueous solution of poly(acrylic acid) acts as a liquid, while powder is composed of properly processed oxides of magnesium and zinc. While application of the mentioned filling, poly(zinc acrylate) is synthesized. That polymer is characterized by gel consistency. Combining this type of cement with the tooth is accompanied by process of chelating of calcium occurring in dentin by carboxyl groups present in the polymer. However, it is worth mentioning that these cements are characterized by a slightly weak strength properties (such as small bending strength in comparison to other commercially available cements as well as the relatively large thickness that hinders adjusting and deposition of filling in the place of application) [16, 17].

Acrylates represent an important group of components of selected cements based on resin. Particularly key role is played by methyl methacrylate. This compound is almost from the half of the twentieth century used for the deposition of various types of prostheses. Widely used are also composite cements that consist of a mixture of methacrylates (or acrylates) with a glass or with a suitably modified silica. Mentioned restorations belong to the group of materials chemo or light cured. On the market of dental products, a significant role is also played by composites modified with polyacids, i.e. compomers that also can contain acrylates [16, 17].

4. Dentures

4.1. Desirable characteristics of poly(methyl methacrylate)

Dental prostheses are defined as an artificial supplement used to complement or to restore missing teeth. These can be made from a variety of materials including metals, ceramics or plastics. Initially, market of dental products was predominated by ceramic restorations. However, with the development of materials science and polymer technology, ceramics has been supplanted by plastics in this field of medicine [16].

The first dentures based on poly(methyl methacrylate) appeared in the 1930s of the twentieth century. Nowadays, the vast majority of dentures is based on poly(methyl methacrylate) or on the mixture of this polymer with copolymers of methyl methacrylate, butyl methacrylate, or propyl acrylate [17–19].

In dental prosthetics, resins based on acrylates are widely used. What is important, these can be used in the form of a powder or liquid. Such compositions also consist of dibutyl phthalate, acetone as well as a variety of stabilizers [19].

Poly(methyl methacrylate) is the most commonly used polymer. It is a result of many desirable characteristics of this compound with a particular emphasis on high mechanical strength, crack and abrasion resistance, acid resistance as well as the ability of easy coverage of this polymer with selected pigment. Furthermore, economic considerations, ease of application, aesthetics as well as optical properties or biocompatibility are undoubtedly important features speaking in favour of the application of this polymer [13, 17].

Resins based on the above-mentioned polymer show a number of advantages whereby they constitute the desired raw material for the preparation of dentures. Therefore, such materials are used to obtain full and removable partial dentures, prostheses for maintaining and micro-dentures [20].

4.2. Modifications of poly(methyl methacrylate)

Poly(methyl methacrylate) is characterized by some imperfections that should be modified or improved. One of the mentioned imperfections of these polymers are undoubtedly mechanical properties. Over 60% of dental prostheses is broken or cracked within the first 3 years of use. The repair process of damaged dental materials is time consuming and costly. It should be noted, however, that the vast majority of users of such materials are elderly people for whom subsequent visits in the dentist office constitute a troublesome duty. Hence, the improvement of the mechanical properties of prostheses running with simultaneous maintaining of other desirable from the point of view of the patient characteristics becomes a real challenge for such realms as materials science and dentistry [13, 21].

An important disadvantage of poly(methyl methacrylate) is a tendency of microorganisms to adhere to its surface which contributes to the microbial growth on the prosthesis made of this material. One of the results of such phenomena is the occurrence of inflammation in the oral cavity. It is also worth noting that dentures based on poly(methyl methacrylate) are characterized by a significantly larger resistance to wear in comparison to ceramic restorations. Hence, many attempts are undertaken in order to improve the resistance which can be achieved, for example, by increasing the cross-linking degree of the polymer or by change of conditions of the polymerization process.

It is possible to distinguish three methods that can solve the problem of PMMA imperfections. These are the

- Replacement of PMMA by other polymers
- Chemical modification of PMMA by grafting with other polymers
- Introduction to the matrix based on PMMA additives of differ origin [22, 23]

Improvement of the physico-chemical properties of acrylates can also be obtained by copolymerization as well as by development of blends or interpenetrating polymer networks based on poly(methyl methacrylate). An impact on the characteristics of the mentioned acrylic resins has undoubtedly addition of various types of organic or inorganic compounds that takes place during process of the polymerization [22].

On the basis of the research, it was found that the introduction to the resin based on bisphenol A glycidyl methacrylate (BIS-GMA) inorganic fillers of micrometric size causes a significant improvement in the strength properties of the material. Moreover, the same effect is reached by addition of inorganic fillers having a nanosize to the matrix based on poly(methyl methacrylate) [24, 25].

An interesting modification of resins based on PMMA is the addition of titanium dioxide. The nanoparticles of this inorganic oxide contribute to the antibacterial properties of the resin. These properties are the result of cytotoxic effects of oxide radicals generated by treating the titanium compound with UV radiation. Furthermore, it was found that the antibacterial activity of TiO_2 is additionally intensified by the presence of metal or metal oxide (e.g. Fe), and therefore, Acosta-Torres et al. carried out a synthesis of the resin based on poly(methyl methacrylate) containing nanoparticles of TiO_2 and Fe_2O_3. Anehosur et al. came to the same conclusion. In the research, they have identified the biocidal properties of the acryl polymer resin modified with titanium dioxide in relation to the selected bacterium, i.e. *Staphylococcus aureus*. Numerous studies and observations conducted by these scientists lead to the conclusion that titanium dioxide subjected to an appropriate radiation effects on the inhibition of the growth and the development of the mentioned strain of bacteria [17, 26, 27].

Next interesting additive that has an impact on mechanical properties of acryl polymer resins is zirconium oxide. Gad et al. on the basis of their research concluded that introduction of the mentioned oxide improves significantly flexural strength of described dentures. Better effect is observed by the introduction into acrylic resin zirconium oxide in the form of nanotubes as evidenced by the conclusions drawn on the basis of the research of Yu et al. [19, 28].

It has also been proved that the addition of substances such as nitrile rubber or materials of ceramic origin such as aluminium oxide affects the improvement of properties such as impact strength, fracture strength or hardness. Such conclusions have been reached by Alhareb et al. [29].

An interesting addition to the acrylic resins forming of the base of dental prostheses represents nanoparticles of gold, silver and platinum. These substances give the mentioned material antifungal properties. This addition is intended to prevent the occurrence of fungal diseases as well as to improve oral hygiene. Promising effect was observed in the case of the introduction of nanoparticles in an amount of 2.0 wt.%. Then, the most visible antiadherent effect was observed in relation to the fungi of the genus *Candida albicans* [30].

Pan et al. on the basis of their research have been stated that the introduction of hydroxyapatite into PMMA-based resins also results in an improvement of mechanical properties of acrylic dentures made of such composite. Hydroxyapatite is an inorganic compound characterized by the properties that are desired in medical applications, i.e. non-toxicity or biocompatibility. Recently, this interesting material constitutes an addition to the polymer matrix that contributes to the improvement of its mechanical properties [31].

Acrylic materials used in dentistry are commonly reinforced by means of different types of fibres wherein special attention should be paid on ceramic, carbon and glass fibres. Their introduction into the material is aimed towards the improvement of the material resistance to severe stress. Kostoulas et al. [32] in the framework of the research described the effect of the addition of glass fibres on the properties of acrylic denture. Similar conclusions have been drawn by Narva [33] based on research undertaken in the framework of the doctoral dissertation concerning strengthening of dental prostheses.

Analysis of the results allows the conclusion that the presence of glass fibres greatly affects the mechanical properties of the tested prostheses. Significantly greater impact resistance of reinforced dental restorations was observed. Furthermore, dentures modified with glass fibers demonstrated more resistance to cracking. What is essential, it was found that the modification using the described fibers has a better impact on the prosthesis than the addition of metal elements [12, 31–33].

Besides glass fibers, the preferred addition to the dental restorations also provides aramid fibres. Raszewski [12] proved that the modification of the denture with this kind of fibres contributes to a clear improvement in mechanical strength of the tested material. Braden et al. [34] in their publication also drew attention to the effect of the addition of the polyethylene fibres to the prosthesis on its properties. These fibers are characterized by the superior properties in comparison to the previously described glass fiber. However, due to the difficulty of combining them with the acrylic resin, it is necessary to use a low-temperature plasma in order to enable synthesis of this composite [12, 34].

Balos et al. [35] in their research undertook the characteristics of acrylic resins modified with nanosilica. An analysis of the obtained materials included defining of mechanical properties such as elasticity or bending strength. The researchers also determined the cytotoxicity of modified materials. Based on the analysis, it was found that for most of the tested prostheses, addition of SiO_2 nanoparticles resulted in an increase of bending strength and in improvement of the modulus of elasticity. It is noteworthy that the amount of added nanosilica is very important because introduction of too much amount causes that material becomes toxic [35].

4.3. Preparation of dentures on the basis of poly(methyl methacrylate)

Poly (methyl methacrylate) constitutes a raw material for the preparation of artificial teeth. However, this polymer to be able to be used for this purpose should be characterized by a high molecular weight and a high cross-linking degree. The starting material consists of PMMA or a copolymer of the mentioned polymer with 1,4-butanediol dimethacrylate or with ethylene glycol dimethacrylate [12].

Crucial meaning for the quality of the denture has a method of polymerization. It is generally carried out at elevated temperature. However, for some substrates that have a tendency towards spontaneous polymerization, process must be carried out at room temperature [12, 36].

Preparation of artificial teeth from the raw material which is PMMA is a multistep process. Initially, 90% of the mixture is represented by monomers, and the remaining 10% constitutes a pigment selected depending on the desired colour of the final product. Depending on the requirements of the final material, it is possible to obtain colourless acrylic mass or mass with the white or pink colours. White acrylic mass is then used to produce dental crowns, inlays or artificial teeth, while the pink one is useful in obtaining products such as impression trays [12, 36].

The next step is subjecting the mixture of trituration process by means of ball mills. Such process depending on the properties of applied reagents takes 24–48 h. Homogeneous mass after trituration process is left in a closed container until reaching the material of gum consistency. Such a mass is then divided into smaller parts which are distributed to the appropriate forms. The moulds are then assembled and subjected to pressing at elevated temperature. At high temperature and under elevated pressure, polymerization process takes place, and final product is obtained. Next, obtained polymer is gradually cooled and formed into its final form by cutting off unpolymerized portions. Subsequently, obtained teeth based on PMMA are affixed to the polymeric plate. An alternative method for the preparation of artificial teeth is a method using the process of injection into the mould at elevated pressure. Molten polymer mixed previously with pigment is subjected to this process [12, 36].

Prepared dental prostheses are characterized by a lack of smell, and furthermore, their sizes can be easily adjusted according to the user [36].

5. Dental adhesives

The primary function of dental adhesives is affixing to the tooth surface or to the elements included in the oral materials such as dentures or dental bridges used in order to restore or to reconstruct a tooth [37].

As in the case of dental prostheses based on poly(methyl methacrylate), dental adhesives made of this polymer are also characterized by slightly poor mechanical properties. From the user's perspective, the most important is proper adhesion of glue to the enamel or dentin (bone tissue from which the core of the tooth is built).

With inadequate adhesion of the adhesive, which results in insufficient adhesion of the prosthesis, or a dental bridge phenomenon of secondary caries can occur, then treatment is considerably less effective. Hence, various attempts aimed at improvement of the properties of dental adhesives based on PMMA as well as at strengthening the existing bonds between the adhesive and the prosthesis are taken.

Solhi carried out a number of syntheses, during which the surface of the nanoclays (pristine sodium montmorillonite) was grafted by poly(methyl methacrylate). Such an obtained system was characterized by a clearly better mechanical properties and exhibited higher adhesion to the dentin [37, 38].

6. Dental impression trays

Acrylic materials represent an important component used in the preparation of individual impression trays. As in the case of above-described cement materials, also materials used for the preparation of impression trays consist of liquid and powder. The fluid is a mixture of methyl methacrylate (90% of a liquid), a suitable catalyst and an oil of mineral origin, while powder consists of inorganic fillers such as talc, chalk or aluminium oxide [11].

The first step in order to obtain an impression element is preparation of plates of a certain size from a previously prepared mixture of liquid and powder. It is essential to choose a suitable proportion of the components. Then, the mixture is prepared in a silicone container. The plate is then introduced into a gypsum mould; elements protruding from the model are cut off. Next, plate is wiped to a smooth surface, and then prepared material is subjected to polymerization process [11].

7. Dental crowns

The term 'prosthetic crown' refers to the permanent restoration that is used to restore the proper tooth crown. Furthermore, such materials are also used to improve conditions of occlusion, to restore teeth using implants as well as for aesthetic purposes.

Among prosthetic crowns, uniform and complex crowns can be distinguished wherein acrylic materials play a significant role in both of these types. The first of the mentioned is made up of one type of material, i.e. acrylates, metal or porcelain. Those made of acrylates in dental prosthetics serve as temporary crowns; however, taking into account the properties of polymer material, which is used predominantly porcelain. The characteristic that causes less and less interest in acrylic materials is undoubtedly low abrasion resistance [11].

8. Studies on physico-chemical properties of prepared dental materials

An important step in the preparation of dental materials is the development of their composition and synthetic parameters. However, an equally important step is performing a series of tests of the obtained materials for characterization of their physico-chemical properties or in order to evaluate whether reagents selected for the synthesis lead to the formation of elements having the desired characteristics. Therefore, it is necessary to carry out durability tests or spectroscopy. Analysis of the results of these studies is important from the point of view of both the user and the dentist.

Figure 7 shows examples of probes applied in the strength studies of prepared dental components; accurate description of the probes is presented in **Table 1**.

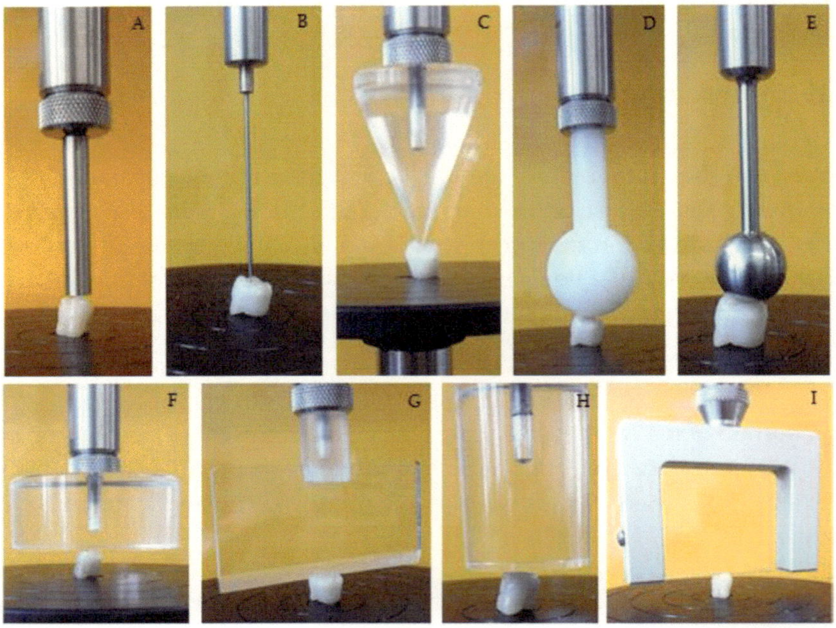

Figure 7. Examples of probes applied in the strength studies.

Designation	Type of probe
A	Cylinder, stainless steel 6 mm
B	Needle, stainless steel 1.0 mm
C	Cone, acrylic glass 40 mm
D	Sphere, nylon 25.4 mm
E	Sphere, stainless steel 12.7 mm
F	Cylinder, acrylic glass 50.8 mm
G	Blade, acrylic glass 60 mm
H	Cylinder, acrylic glass 25 mm
I	Wire, stainless steel 0.03 mm

Table 1. Set of probes applied for strength tests.

In **Figure 8** examination of synthesized artificial tooth using a spectrophotometer is shown.

Above-presented studies are extremely important from a practical point of view and in terms of the subsequent application. Physical and chemical properties of dental materials described in this section have an impact on the comfort of future use and on maintaining of the oral health of the patient. Therefore, carrying out various types of mechanical studies and analysis of the attained results in order to determine the final properties of the prepared product is necessary before implementation.

Figure 8. Spectroscopic analysis of prepared dental materials.

Author details

Bożena Tyliszczak[1]*, Anna Drabczyk[2] and Sonia Kudłacik-Kramarczyk[2]

*Address all correspondence to: btyliszczak@chemia.pk.edu.pl

1 Department of Chemistry and Technology of Polymers, Cracow University of Technology, Krakow, Poland

2 Institute of Inorganic Chemistry and Technology, Cracow University of Technology, Krakow, Poland

References

[1] Ilewicz L. Materiały do wypełnień we współczesnej dentystyce odtwórczej. 1st ed. Bielsko-Biała; 2003. 108 p

[2] Coleman RL. Physical properties of dental materials. Journal of Research of the National Bureau of Standards. 1928;**15**(6):1053-1054

[3] Craig RG, Farah JW. Stress analysis and design of single restorations and fixed bridges. Journal of Applied Oral Science. 1977;**10**:45-74

[4] Craig RG. Advances in biomaterials from 1957-1997. Journal of Oral Rehabilitation. 1999;**26**(11):841-846

[5] Osborne J. Acrylic Resins in Dentystry. 3rd ed. London; 1949. 100 p

[6] Smith DC. The genesis and evolution of acrylic bone cement. Orthopedic Clinics of North America. 2005;**36**(1):1-10

[7] Price CA. History of dental polymers. Australian Prosthodont Journal. 1994;**8**:47-54

[8] Murray MD, Darvell BW. The evolution of the complete denture base. Theories of complete denture retention – A review, Part 1. Australian Dental Journal. 1993;**38**(3):216-219

[9] Hargreaves AS. The development of denture design. British Dental Journal. 1981;**151**(3): 101-102

[10] Peyton FA. History of resins in dentistry. Dental Clinics of North America. 1975;**19**(2): 211-222

[11] Craig RG . Materiały stomatologiczne. 12th ed. Wroclaw: Elsevier Urban & Partner; 2008. 603 p

[12] Raszewski Z. Nowe spojrzenie na tworzywa akrylowe. 1st ed. Katowice: Elamed; 2009. 182 p

[13] Gurbuz O, Unalan F, Dikbas I. Comparison of the transverse strength of six acrylic denture resins. OHDMBSC. 2010;**9**(1):21-24

[14] Niewczas A, Pieniak D, Bachanek T, Surowska B, Bienias J, Palka K. Prognozowanie degradacji czynnościowej systemów biomechanicznych na przykładzie układu ząb - wypełnienie kompozytowe. Eksploatacja i Niezawodność. 2010;**1**:23-34

[15] Rošin-Grget K, Peroš K, Šutej I, Bašić K. The cariostatic mechanisms of fluoride. Fluoride and Caries Prevention. 2013;**42**(2):179-188. DOI: 10.5644/ama2006-124.85

[16] Szafran M, Rokicki G, Czerwińska L, Bobryk E. Zastosowanie rozgałęzionych żywic uretanowo-metakrylowych w kompozytach stomatologicznych. Kompozyty. 2008;**8**(1):47-52

[17] Acosta-Torres LS, Lopez-Marin LM, Nunez-Anita RE, Hernandez-Padron G, Castano VM. Biocompatible metal-oxide nanoparticles: Nanotechnology improvement of conventional prosthetic acrylic resins. Journal of Nanomaterials. 2011;**2011**:1-8. DOI: 10.1155/2011/941561

[18] Anusavice KJ, editor. Phillips' Science of Dental Materials. 11th ed. USA: Elsevier Health Sciences; 2003. 832 p

[19] Yu W, Wang X, Tang Q, Guo M, Zhao J. Reinforcement of denture base PMMA with ZrO_2 nanotubes. Journal of the Mechanical Behavior of Biomedical Materials. 2014;**32**:192-197. DOI: 10.1016/j.jmbbm.2014.01.003

[20] Körholz K-H, Mikuš M. Rewolucja w tworzywach akrylowych. Nowa generacja sztucznych żywic na bazie PMMA. Nowoczesny Technik Dentystyczny. 2009;**2**:37-44

[21] Yamauchi M, Yamamoto K, Wakabayashi M, Kawano J. In vitro adherence of microorganisms to denture base resin with different surface texture. Dental Materials Journal. 1990;**9**(1):19-24

[22] Reis KR, Bonfante G, Pegoraro LF, Conti PCR, Oliveira de PCG, Kaizer OB. In vitro wear resistance of three types of polymethyl methacrylate denture teeth. Journal of Applied Oral Science. 2008;**16**(3):176-180. DOI: 10.1007/s00784-015-1502-8

[23] Jagger DC, Harrison A, Jandt KD. Review: The reinforcement of dentures. Journal of Oral Rehabilitation. 1999;**26**:185-194. DOI: 10.1046/j.1365-2842.1999.00375.x

[24] Whitman DJ, McKinney JE, Hinman RW, Hesby RA, Pelleu GB. In vitro wear rates of three types of commercial denture tooth materials. Journal of Prosthetic Dentistry. 1987;**57**:243-246. DOI: 10.1590/S1678-77572008000300003

[25] Suzuki S. In vitro wear of nano-composite denture teeth. Journal of Prosthodontics. 2004;**13**:238-243. DOI: 10.1111/j.1532-849X.2004.04043.x

[26] Li Q, Mahendra S, Lyon DY et al. Antimicrobial nanomaterials for water disinfection and microbial control: Potential applications and implications. Water Research. 2008;**42**(18): 4591-4602

[27] Anehosur GV, Kulkarni RD, Naik MG, Nadiger RK. Synthesis and determination of antimicrobial activity of visible light activated TiO_2 nanoparticles with polymethyl methacrylate denture base resin against *Staphylococcus aureus*. Journal of Gerontology & Geriatric Research. 2012;**1**(1):103-111. DOI: 10.4172/2167-7182.1000103

[28] Gad MM, Rahoma A, Al-Thobity AM, ArRejaie AS. Influence of incorporation of ZrO_2 nanoparticles on the repair strength of polymethyl methacrylate denture bases. International Journal of Nanomedicine. 2016;**27**(11):5633-5643. DOI: 10.2147/IJN.S120054

[29] Alhareb AO, Akil HM, Ahmad ZA. Impact strength, fracture toughness and hardness improvement of PMMA denture base through addition of nitrile rubber/ceramic fillers. The Saudi Journal for Dental Research. 2017;**8**(1-2):26-34. DOI: 10.1016/j.sjdr.2016.04.004

[30] Nam KY. Characterization and antifungal activity of the modified PMMA denture base acrylic: Nanocomposites impregnated with gold, platinum, and silver nanoparticles. In: Grumezescu AM, editor. Nanobiomaterials in Dentistry, Applications of Nanobiomaterials. UK: Matthew Deans; 2016. pp. 309-336. DOI: 10.1016/B978-0-323-42867-5.00012-6

[31] Pan Y, Liu F, Xu D, Jiang X, Yu H, Zhu M. Novel acrylic resin denture base with enhanced mechanical properties by the incorporation of PMMA-modified hydroxyapatite. Progress in Natural Science: Materials International. 2013;**23**(1):89-93. DOI: 10.1016/j.pnsc.2013.01.016

[32] Kostoulas I, Kavoura VT, Frangou MJ, Polyzois GL. Fracture force, deflection, and toughness of acrylic denture repairs involving glass fiber reinforcement. Journal of Prosthodontics. 2008;**17**(4):257-261. DOI: 10.1111/j.1532-849X.2007.00276.x

[33] Narva K. University of Turku Academic Dissertation. Doctoral Thesis on Reinforcing Denture with Fiber Reinforcements [dissertation]. Turku, Finland: University of Turku; 2004. 630 p

[34] Braden M, Davy KMW, Parker S, Ladizesky HN, Ward IM. Denture base poly(methyl methacrylate) reinforced with ultra-high modulus polyethylene fibers. British Dental Journal. 1988;**164**:109-113

[35] Balos S, Puskar T, Potran M, Markovic D, Pilic B, Pavlicevic J, Kojic V. Modulus of elasticity, flexural strength and biocompatibility of poly(methyl methacrylate) resins with low addition of nanosilica. Research & Reviews: Journal of Dental Sciences. 2016;**4**(1):26-33

[36] Olędzka E, Sobczak M, Kołodziejski WL. Polimery w medycynie – przegląd dotychczasowych osiągnięć. Polimery. 2007;**52**(11-12):795-803

[37] Krifka S, Börzsönyi A, Koch A, Hiller KA, Schmalz G, Friedl KH. Bond strength of adhesive systems to dentin and enamel – Human vs. bovine primary teeth in vitro. Dental Materials. 2008;**24**(7):888-894. DOI: 10.1016/j.dental.2007.11.003

[38] Solhi L, Atai M, Nodehi A, Imani M. A novel dentin bonding system containing poly(methacrylic acid) grafted nanoclay: Synthesis, characterization and properties. Dental Materials. 2012;**28**(10):1041-1051. DOI: 10.1016/j.dental.2012.06.004

Chapter 3

New Trends for the Processing of Poly(Methyl Methacrylate) Biomaterial for Dental Prosthodontics

Ma. Concepción Arenas-Arrocena,
Liliana Argueta-Figueroa, René García-Contreras,
Omar Martínez-Arenas, Berenice Camacho-Flores,
María del Pilar Rodriguez-Torres,
Javier de la Fuente-Hernández and
Laura S. Acosta-Torres

Additional information is available at the end of the chapter

http://dx.doi.org/10.5772/intechopen.69066

Abstract

Rehabilitation of masticatory function in patients with absent teeth with removable dentures is an established form of treating partial or complete dentition in edentulous patients. The developments in recent decades with dental implants dominate current dental research. However, medical contraindications, a negative attitude toward implants, or financial limitations on the part of the patients limit their universal applicability, so the rehabilitation with dental prostheses still makes up a significant portion of everyday clinical practice. Conversely, removable dentures are used in the critical conditions of the oral cavity. There are about 500 strains of microorganisms in the mouth, which form the biofilm in an acidic environment causing several issues, such as denture stomatitis, deterioration of the periodontal status of the remaining teeth, or carious lesions in the supporting teeth. Therefore, it is very important to choose a suitable material for the prosthesis. Poly(methyl methacrylate) (PMMA) is an acrylic resin usually used with a long tradition for prosthetic purposes. The aim of this chapter is to present the trends for the processing of PMMA. It includes the chemical synthesis, conventional thermal processing of this acrylic resin, the new processing technique assisted with ultrasound, the antibacterial effect on PMMA with nanoparticles, and the cytotoxicity, genotoxicity, and mutagenesis of this material.

Keywords: thermal polymerization, acrylic resin, biomaterial, polymer, dental materials, acrylic resin

1. Introduction

The dynamic development of new multidisciplinary areas has a direct impact over the possible treatments and the rehabilitation of the dental function. Teeth rehabilitation with removable denture prosthesis is an established form of treating both partial and complete dentition in edentulous patients [1]. The developments in recent decades with dental implants dominate the current dental research, not only medical contraindications but also a negative attitude toward implants [2] and economic limitation [3] are the major disadvantages for their universal applicability, so the rehabilitation with dental prostheses still makes up a significant portion of everyday clinical practice [4].

The PMMA material revolutionized the preparation techniques used so far since Walter Wright introduced the acrylic resin as the denture base material in 1937 [5]. The acrylic resin became the preferred material for making denture bases, due to its ability to overcome many of the deficiencies of the materials used at that time [6].

Conversely, removable dentures are used in critical conditions of the oral cavity. There are about 500 microorganisms in the mouth, which produce a biofilm in an acidic environment causing several diseases [7], such as denture stomatitis [8], deterioration of the periodontal status of the remaining teeth [9], or carious lesions in abutment teeth [10]. Therefore, it is very important to choose a suitable material for dental prosthesis.

Poly(methyl methacrylate) (PMMA) is an acrylic resin usually used with a long tradition for prosthetic purposes [11]. It can be classified as chemically or thermally polymerized material depending on the factors that initiate the reaction. For dental prosthesis, thermally polymerized materials are used and the heat can be generated by hot water bath or microwave energy [12]. It was suggested that residual monomer concentration is the most important parameter in the determination of the final properties of the PMMA for dental prosthesis [12, 13]. It was found that in the chemical structure of PMMA, the alpha methyl groups tend to remain in the outer layer surface, whereas the methylene groups are in the inner layer of the PMMA surface, which gives an idea of the arrangement of the polymer [13]. In other words, PMMA has exhibited moderate cytotoxicity in bulk material and polymerized form [14, 15].

The aim of this chapter is to present the trends for the processing of PMMA, including the chemical synthesis, conventional processing (thermal polymerization), the new technique of thermal polymerization assisted with ultrasound, the antibacterial effect on PMMA with nanoparticles, and biocompatibility (cytotoxicity, genotoxicity, and mutagenesis).

2. Poly(methyl methacrylate) (PMMA): synthesis, morphology, and physical properties

Acrylic acid ($C_3H_4O_2$) gives rise to the so-called acrylic, where the poly(methyl methacrylate) (PMMA) is the most important thermoplastic in this group, which is commercially known as Plexiglas, Lucite, and Perspex [16].

PMMA is an amorphous polymer formed by the polymerization of MMA monomer carried out using different mechanisms [free radical vinyl polymerization, anionic polymerization, group transfer polymerization (GTP), or atom transfer radical polymerization (ATRP)] [16–20]. The bulk or solution (homogeneous polymerization) and emulsion or suspension (heterogeneous polymerization) techniques are used to obtain PMMA [18, 20–22]. Among them, suspension polymerization is a good route to produce PMMA with high molecular weight (36,100), high yield (83%), and a polydispersity of 2.4 (polydispersity index: Mw/Mn) [18].

2.1. Suspension polymerization

Hoffman and Delbruch developed suspension polymerization in 1909 for the first time [23]. In this technique, the initiator and the monomer are miscible with each other (**Figure 1**) and it involves droplet formation by the initiator/monomer (polymerizing phase) dispersed into water (oil/water system), where the volume ratio of monomer about 0.5 or less is suggested [22]. Water works as a heat-transfer agent and a dispersion medium, which improves the reaction rate and the yield in the polymerizing phase. To prevent settling or creaming, the suspension polymerization was kept under stirring during polymerization. In this polymerization, the addition of a soluble stabilizer in water [gelatin, clay or clay derivative, cellulose derivatives, water-soluble polymers such as poly(vinyl alcohol) (PVA) or starch] helps to prevent the breakup of droplets or avoids the droplet from adhering to each other [20–22, 24]. This process could be assisted with low temperature or ultrasonic waves [20, 24–27].

2.1.1. Spherical microparticles: effect of stabilizer agent on size

Suspension polymerization is adequate technically to obtain PMMA spherical microparticles with controlled sizes ranging from 5 to 1000 μm [22]. Alginate stabilizer produces microparticles from 5 to 80 μm (**Figure 2a**), whereas microparticles below 30 μm are obtained with gelatin stabilizer (**Figure 2b**) as previously reported [24–26]. These sizes are within the range of commercial PMMA (10 to 100 μm) used for prosthodontics (**Figure 2c** and **d**). Therefore, polydisperse particles could influence the surface roughness of PMMA without affecting their mechanical properties [28].

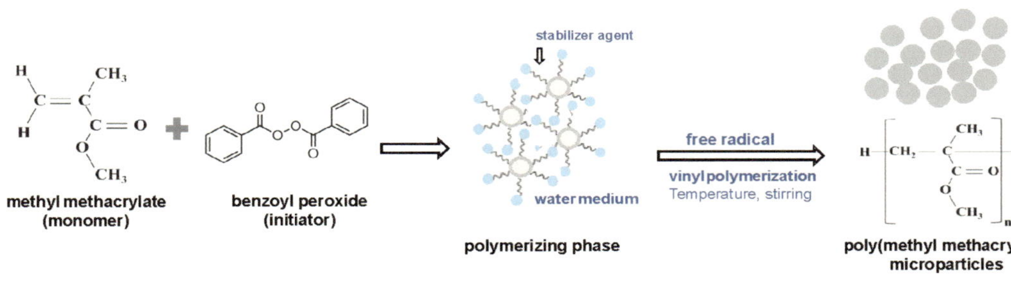

Figure 1. Suspension polymerization of MMA monomer. In the first step, the initiator, benzoyl peroxide interacts with the monomer in water in order to form emulsion oil/water, where the volume of water is twice as that of the monomer. Soluble water-stabilizer helps to obtain smooth and controlled size spherical microparticles.

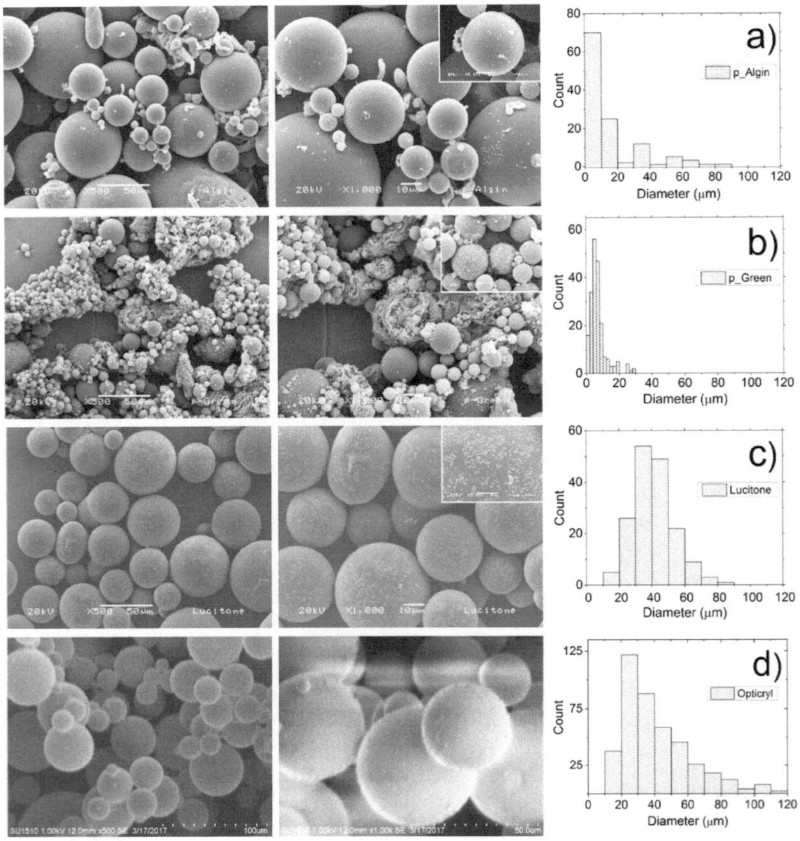

Figure 2. Experimental PMMA microparticles obtained by suspension polymerization with (a) alginate or (b) gelatin stabilizer agents. Commercial PMMA microparticles: (c) Opticryl®, (d) Lucitone® used for prosthodontics.

2.2. Physical properties

PMMA has different characteristics and properties, such as chemical stability, hardness, stiffness and high transparency, resistance in atmospheric conditions and greater impact resistance than glass, and thermal and acoustic insulation. **Table 1** enlists the physical properties of PMMA [17, 29].

Properties	Values
Relative molecular mass	100.12
Elastic modulus	2.4–3.1 GPa
Tensile strength	80 MPa
Flexural strength	140 MPa
Elongation at break	2–5%
Volatility	3.87 kPa at 20°C
Stability	Highly inflammable vapor, lower explosive limit 2.1 vol%

Properties	Values
Glass transition temperature (T_g)	100–130°C
Fusion temperature	200°C
Density	1.2 g/cm^3
Refractive index	1.49
Water absorption	30 mg

Table 1. Properties of poly(methyl methacrylate) [17, 29].

These properties are important for the final application, such as optical device, airplane windows, lenses, covers, automotive taillight, dental articles, and bioengineering [29]. Also, PMMA is a material widely used daily in dental practice, such as dental prosthesis for edentulous patients [24, 26]. For this particular application, PMMA (experimental or commercial acrylic resin) must be processed by heat, which can be generated by hot water bath or microwave energy [12].

3. Thermal polymerization processing of PMMA

The denture bases made up of acrylic PMMA resin, which is in contact with the oral mucosa of the patient is a critical aspect for biocompatibility in contact with tissues. The PMMA resin was chosen due to this important adequate processing technique [14]. Polymerization of PMMA by water bath and microwaves are the most commonly used processing techniques for making denture bases [12]. The water bath and microwave polymerization techniques produce a material with reduced porosity and irregularities on the PMMA surface. Independent of the processing method, the PMMA surface exhibits some defects (pores, cracks, and irregularities) that are produced at the time of its elaboration [30, 31]. These defects can be excellent reservoirs for fungi and opportunistic bacteria, besides decreasing the elastic modulus and flexural strength [12, 13, 32].

Over the years, the water bath processing technique has been the most widely used due to its ease of handling and cost effectiveness. But, the residual monomer content and porosity have been suggested as the most significant reasons for the reduced flexural strength [33]. It has been accounted the unfavorable thermal gradient produced during the processing technique. In the water bath processing technique, the benzoyl peroxide (initiator) was activated by heating the water to a very high temperature, which leads the polymerization reaction by crosslinking methyl methacrylate moieties. At this point, the methyl methacrylate particles begin to boil by creating porosities in the denture base resin [34]. As the reaction progresses, heat is liberated and cannot escape easily as the water surrounding the flasks is being heated as well. Thus, an unfavorable thermal gradient was created [35].The residual monomer inside the polymeric mass can negatively influence the physical and mechanical properties of the materials due to its plasticizing action [36]. On the other hand, during the microwave polymerization, monomer molecules move in a high-frequency electromagnetic field [37]. The microwaves cause the methyl methacrylate molecules within the acrylic resin to orient themselves in the electromagnetic field at a frequency of 2450 MHz [38], and numerous polarized molecules are flipped over rapidly and generate heat due to molecular friction [39]. Numerous intermolecular collisions are promoted, causing a rapid internal heating in which energy was immediately absorbed by the resin regardless of the thermal conductivity of the materials involved in the processing of the prosthesis [40]. This warming occurs rapidly and homogeneously and thereby transfer of heat from the water bath to the resin inside the flask occurs faster in this method [41].

There are several studies to compare the flexural strength and elastic modulus values of PMMA using water bath and microwave polymerization [13, 35, 36, 41–43]. In most cases, the results of microwave polymerization did not differ from those obtained with water bath, independent of the acrylic resins used [41, 43]. However, in some studies, water bath technique showed higher flexural strength than microwave processing technique [44]. On the contrary, in other studies, a statistically higher flexural strength was found for microwave-processed denture resins [45, 46]. Other researchers did not find a significant difference in porosity between microwave polymerization and conventional water bath cycles [39, 47, 48]. In contrast, the other work reported that heat polymerization technique presents lower mean porosity values than microwave-polymerization method [31]. Both processing techniques produced PMMA material with divergent properties. Therefore, new processing techniques for PMMA are needed to reduce the amount of residual monomer and porosity and to increase its physical strength.

3.1. Thermal polymerization assisted with ultrasound

The most widely used heat-curable acrylic material to make dental bases and temporary restorations is PMMA. A disadvantage of this acrylic resin is the residual monomer which remains in the polymer even after its polymerization is finished [49]. Several attempts were made to find a better strategy in order to prevent the presence of residual monomer. For example, the effects of temperature, time, initiator concentration, curing environment, water bath or microwave oven, pressure, and mixing ratio (polymer:monomer) have been investigated [32].

The first effort to employ ultrasound for the acceleration of conventional chemical reactions [50] by Richards and Loomis was reported in 1927. A lot of interest has been attracted for the use of ultrasound toward the development of synthetic routes in a variety of areas of chemistry, chemical production, and materials science [27, 51]. It is possible to generate chemical changes in consequence of acoustic cavitation while more powerful ultrasound at a lower frequency is applied to a system. During cavitation, bubble collapse produces intense local heating, high pressures, and very short lifetimes. These transient and localized hot spots drive high energy toward completing chemical reactions faster [52]. Besides, the physical effect of the medium on the wave was referred to low power or high frequency ultrasound [53].

In previous studies, Charasseangpaisarn and Wiwatwarrapan [49, 54] found that the use of an ultrasonic treatment at several frequencies reduced the presence of residual monomer in acrylic resins. For example, heat-polymerized MMA by the immersion in water at 50°C for 10 min at 40 kHz reduced the residual monomer. They have concluded that sonication could reduce the amount of residual monomer in acrylic resins. According to the authors, the ultrasonic treatment could enhance the extraction rate of the residual monomer from the resin and could cause postpolymerization of the residual monomer.

3.1.1. *Influence of frequency and power of ultrasonic waves on the flexural strength and elastic modulus*

The method of denture processing is directly related to the physical properties of the acrylic resins. One of those properties is Young's modulus, also known as elastic modulus. That is defined as capacity of a body to deform to the application of stress and strain after removing the body recovers its original shape. It can be assumed that the relationship between the increased effort and increased deformation is constant [55]. Flexural failure of denture base of PMMA is considered to be the main form of clinical failure [56]. The dental prostheses are subjected to various conditions such as forces during chewing, drastic changes of temperature and humidity, and acidic environment in the oral cavity. Therefore, it is important that prosthetic materials possess an adequate elastic modulus [42]. The elastic modulus can be determined by indentation techniques. However, the correct use of these techniques requires knowing their limitations in order to avoid misinterpretation.

Experimental results about the elastic modulus and flexural strength (ISO20795-1:2008 *Part 1: Denture base polymers*) of commercial acrylic resin (Opticryl®) indicate that the thermo-polymerization assisted with ultrasound is a good option for the processing of PMMA. Commercial acrylic resin (Opticryl®) specimens (n = 25) were prepared according to the technical sheet with a volume ratio of monomer to polymer (1:6). For the processing condition by ultrasound waves, two frequencies and powers were used at 80°C of water bath for 1 hour: 37 or 80 kHz and 50 or 100%, respectively, in order to obtain four experimental groups (**Table 2**). Water bath and microwave technical processing were considered to be the control groups. The results of the elastic modulus and flexural strengths are given in **Figure 3** and **Table 2**.

For statistic comparison among the groups, Kruskal-Wallis test and Mann-Whitney U-test were used for analyzing the data. These tests were used because not all groups had a normal distribution as shown by the Shapiro-Wilk normality test (see **Table 3**).

Kruskal-Wallis test showed that there are significant differences among the groups (for elastic modulus, $p = 0.006$ and for flexural strength, $p = 0.018$). Mann-Whitney U test was conducted

Group	Thermopolymerization	Frequency (kHz)/ power (%)	Elastic modulus (MPa)	Flexural strength (MPa)
1	Ultrasonic	37/50	1710.38 ± 429	58.63 ± 13.8
2	Ultrasonic	37/100	1730.75 ± 335.13	58.16 ± 10.64
3	Ultrasonic	80/50	1488.86 ± 80.02	51.05 ± 9.61
4	**Ultrasonic**	**80/100**	**1828.08 ± 363.67**	**62.14 ± 12.92**
5	Water bath	–	1744.40 ± 441.85	60.57 ± 14.91
6	Microwave	–	1466.12 ± 428.39	54.15 ± 17.13

Specimens processed by water bath and microwaves are considered the control groups.

Table 2. Elastic modulus and flexural strength of specimens processed by ultrasound at 80°C: 37 or 80 kHz and 50 or 100% of power under constant temperature of water (80°C).

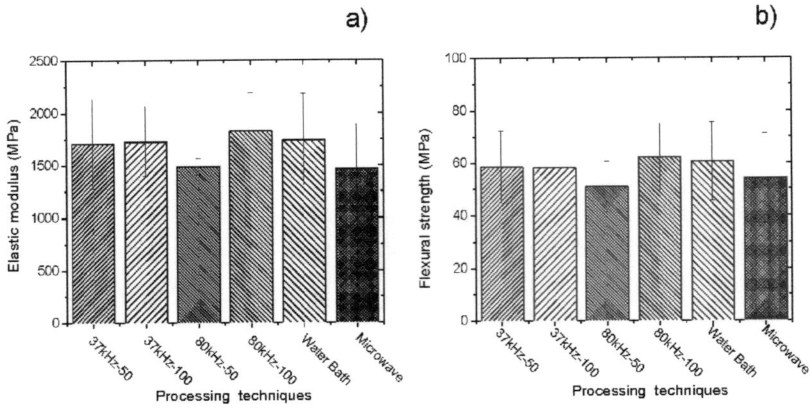

Figure 3. Results of elastic modulus and flexural strength of commercial Opticryl resin polymerized by ultrasound. Control groups were processed by water bath and microwave energy.

Mechanical properties	Frequency (kHz)/ power (%)	Shapiro-Wilk statistic analysis		
		Statistic value	Degrees of freedom	*p*-Value
Elastic modulus	37/50	0.940	32	**0.076***
	37/100	0964	38	**0.261***
	80/50	0.844	26	0.001
	80/100	0.922	35	0.017
	Water Bath	0.967	19	**0.711***
	Microwave	0888	21	0.020
Flexural strength	37/50	0.893	32	0.004
	37/100	0.962	38	**0.225***
	80/50	0.924	26	0.056*
	80/100	0.932	35	0.032
	Water bath	0.955	19	**0.472***
	Microwave	0.919	21	**0.083***

*The data follow a normal distribution if $p \geq 0.05$.

Table 3. Results of Shapiro-Wilk normality test.

among groups in all possible combinations to determine the differences among groups. The results are shown in **Table 4**.

The specimens processed at 80 kHz and 100% of power (group 4) exhibited the highest values with an elastic modulus of 1744.40 ± 441.85 MPa and a flexural strength of 60.57 ± 14.91 MPa. However, the flexural strength values were not statistically significant compared to those processed by the water bath and microwave, respectively (**Figure 3**). But, with regards to

Comparison among groups	Elastic modulus		Flexural strength	
	Value of the test statistic	*p*-Value	Value of the test statistic	*p*-Value
1–2	563,000	0.596	535,000	0.389
1–3	321,000	**0.137***	269.500	**0.022***
1–4	439,000	0.129	452,500	0.177
1–5	276,000	0.585	274,000	0.559
1–6	242,000	0.087	287,000	0.373
2–3	322,000	**0.019***	273,000	**0.003***
2–4	563,000	0.260	579,000	0.342
2–5	341,000	0.735	310,000	0.388
2–6	255,000	**0.023***	363,000	0.569
3–4	227,000	**0.001***	213,000	**0.001***
3–5	163,000	**0.054***	153,000	**0.031***
3–6	262,000	0.814	252,000	0.653
4–5	312,000	0.710	322,000	0.849
4–6	187,000	**0.002***	284,000	0.158
5–6	130.000	0.60	152,000	0.198

*Statistical significance, $p \leq 0.05$.

Table 4. Comparison among groups using Mann-Whitney *U*-test.

the elastic modulus, a highly significant difference among the specimens of group 4 and the specimens processed with microwave (group 6) was found. Hence, these results indicate a better performance of the PMMA processed by ultrasound in 80 kHz and 100% of power in comparison to that by microwave processing.

In addition, it is further noted that the acrylic resins processed at 80 kHz and 50% of power (group 3) had significantly lower values compared to the other experimental groups, for both the elastic modulus and the flexural strength. No statistically significant differences in the elastic modulus and flexural strength among groups (water bath and microwave) were found. Therefore, it was concluded that these two methods have similar results. From these results, it seems that the power is more important than the frequency of ultrasound for better results in the processing of PMMA.

Spearman correlation test was performed in order to determine if the elastic modulus values and flexural strength values are correlated. It was found that a weak correlation existed since the correlation coefficient between the elastic modulus and flexural strength was 0.618 ($p \leq 0.001$). Since the coefficient is a positive value, the increasing elasticity modulus value also increases the flexural strength.

In summary, the best conditions for higher values of both elastic modulus and flexural strength correspond to the specimens processed at 80 kHz and 100% of power (group 4). The processing of PMMA with water bath or microwave processing generated similar values for elastic modulus and flexural strength. Ultrasound can be used to process the acrylic resin (Opticryl®) as an alternative technique for PMMA processing with similar results to those obtained using water bath or microwave processing (control groups). The correlation coefficient between the elastic modulus and the flexural strength indicates a weak correlation but statistically significant association between these two variables. The sign of the coefficient is positive, this means that as the values of the elastic modulus increase, those of the flexural strength also increase.

4. Biological properties of PMMA

4.1. Antimicrobial activity

As mentioned above, the current techniques for processing base denture produce porosities, which allow bacterial colonization [30, 31]. One way to approach this issue is the covering up of the PMMA surface. Since the introduction of nanoparticle-based antimicrobial agents, these have generated really huge interest. Diverse mechanisms for explaining the activity of antimicrobial agents have been discussed, such as the release of ions from the nanoparticle surface, the internalization through cell wall, the production of reactive oxygen species [57], and the destruction of cell wall by the nanometric pillars on the surfaces, among others [58]. For instance, the wing surfaces of insects such as dragonflies and cicadas exhibit a texture that is formed by nanopillars, which are very effective against certain type of pathogenic microorganisms [58, 59]. The possibility of developing surfaces that have antibacterial effects quickly became the subject of study [60].

The characteristics of the surfaces of certain objects make them excellent places for proliferation of pathogenic microorganisms and thereby prevent the bacterial adhesion. The main characteristics of polymer surfaces related to microbial adhesion are chemical composition and topography [61].

Different surface modifications have been suggested to reduce the adhesion of pathogenic microorganisms. At present, one of the most effective methods is the surface modification with metallic antibacterial agents such as silver, copper, and zinc oxide at nanometric scale [62, 63]. It has been demonstrated that the oxidized state on surfaces (through electrochemical anodization) shows a significant decrease of some bacterial strains present in the oral cavity and bacteria involved in the process of biofilm formation [64].

Besides, polymeric glycol-based coatings have been proposed in order to immobilize the molecules on the surface of the substrate. Thus, this prevents bacterial adhesion. The modification of the surface topography generates an unfavorable surface chemistry for the adhesion of certain microorganisms and therefore the colonization of surfaces [65, 66].

The arrangement of polymeric coatings with antibacterial agents such as nanoparticles has been studied. The best alternative is that the nanoparticles have to be contained in the polymer matrix, so that their release acts at the level of biofilm formation [67]. According to the type of antimicrobial agent and disposition on the surface, it may offer more than one function eradicating an acute infection and even providing extended periods of suppression of bacterial proliferation [68, 69]. The action can be differentiated depending upon its mechanism in passive coatings (coatings that prevent bacterial adhesion), contact-killing coatings, and active coatings with the ability to release the antibacterial agent incorporated [70]. A coating includes different antimicrobial agents, such as moieties, nanoparticles, and antibiotics for specific pathogens [66, 71]. Silver nanoparticles as a cover on PMMA decrease the roughness from 566.7 nm (without nanoparticles) to 104.08 nm (with nanoparticles) (**Figure 4**). On the other hand, Ziad *et al.* found that Nystatin modifies the roughness of PMMA so that this could influence the antifungal agents on the PMMA surface [66].

These results show that PMMA with antimicrobial agent are potentially useful for their application in dentures for the future. Not many studies have been carried out and there is still scope for further study in this area.

In addition to the antibacterial effect, PMMA-metal oxide nanoparticles have been synthesized with the purpose of improving PMMA's flexural strength as well [63]. With this aim, several works have been carried out by incorporating TiO_2 nanoparticles and assessing the dependence of the flexural strength on the TiO_2 nanoparticle concentration. It was observed that by increasing the concentration of nanoparticles, the flexural strength of PMMA value increases. In some cases, better flexural strength value was found in comparison with PMMA alone [72, 73]. Studies on the improvement of tensile strength concluded that increasing the TiO_2 nanoparticle concentration provided better tensile strength up to some concentration and then the strength decreases [74]. Recently, Totu *et al.* developed a PMMA-TiO_2 material with improved antibacterial activity, for manufacturing 3D-printed dental prosthesis [75].

Other metal oxide nanoparticles that have also been used for their integration to PMMA are the iron dioxide nanoparticles [25]. These nanoparticles improved the antimicrobial and mechanical properties of the acrylic resins. Nanopigmented particles incorporated into PMMA also have been shown to be non-cytotoxic (against fibroblast *in vitro*) and to exhibit good physical

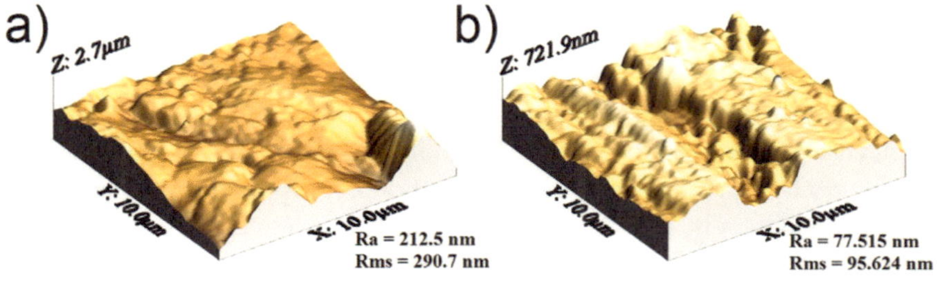

Figure 4. Surface roughness of (a) PMMA uncoated and (b) coated with silver nanoparticles by spin coating.

and mechanical properties as well [24]. In both the cases, specimens exhibited good mechanical and physical properties and were not non-cytotoxic showing similar appearance to commercial acrylic resins.

4.2. Biocompatibility

One of the most important factors that distinguish biomaterials is its ability to exist into or in contact with tissues of the human body without inducing any collateral effect, where both biomaterials and tissues coexist, and the biocompatibility may be compromised.

Biocompatibility refers to the ability of a material to perform with an appropriate host response in a specific situation [76].

The biocompatibility of a material depends on the type of material, where it is placed, and the function it is expected to perform. Therefore, a biocompatible material elicits an acceptable tissue response when tested or used in a specific tissue under certain conditions, including the health status of the patient [77]. It is important to understand the paradigms of biocompatibility by the determination of which chemical, biochemical, physiological, physical, or other mechanisms, under specific conditions, associated with contact between biomaterials and cells or tissues of the body. The interactions of materials that are in direct contact with the human body depends on the characteristics of the host such as age, sex, general health and current disease, physical mobility, lifestyle features, and pharmacological status [78]. Thus, the major features influenced in the host and generic host response (implanted or in contact with tissues) of biomaterials are enlisted in **Table 5**.

On the other hand, PMMA-based acrylic resin has been broadly used as a dental material, especially in denture base processing due to its favorable working characteristics, processing ease, accurate fit, stability in the oral environment, and superior aesthetics with inexpensive equipment. Despite these excellent properties, there is a need for improvement in the biological aspects of biocompatibility. This section is oriented to summarize the different methods of PMMA biocompatibility alone and enriched or modified with different biomaterials in contact with cells and implantation in animal bodies highlighting the type of cells or animal test, period of incubation or implantation, method for analysis, and results. The incorporated studies are recent publications indexed at MEDLINE/PUBMED based on a systematic review.

4.2.1. Test methods

Testing for cytocompatibility depends on the site of use and the duration of exposure. Biomaterials or other associated products do not have to exhibit the same compatibility as materials that are placed permanently into the tooth structure, used as implants into bone or soft tissues, or used in dentures and dental or orthodontic appliances. All the tests are usually conducted sequentially, with shorter term, *in vitro*, or less expensive screening testing, and involve the use of animals. If a material is not showing biocompatibility based on initial studies, it may be better to eliminate it from consideration for further testing for certain applications [79]. The biocompatibility concerns and the testing methods have been discussed for over 40 years. However, new issues and new testing possibilities must be considered for innovating dental materials and evaluate the response of cells to medical materials at the cellular

Variables that could influence the host response	Characteristics of the generic host response to biomaterials
Bulk material composition, micro- (or nano)-structure, morphology	Protein adsorption and sorption characteristics
Crystallinity and crystallography	General cytotoxic effects
Elastic constants	Neutrophil activation
Water content, hydrophobic–hydrophilic balance	Macrophage activation, foreign body giant cell production, granulation tissue formation
Macro-, micro-, nano-porosity	Fibroblast behavior and fibrosis
Surface chemical composition, chemical gradients, surface molecular mobility	Microvascular changes
Surface topography and energy	Tissue/organ-specific cell responses (e.g., osteoclasts and osteoblasts for bone, endothelial proliferation)
Surface electrical/electronic properties	Activation of clotting cascade
Corrosion parameters, ion release profile, metal ion toxicity (for metallic materials)	Platelet adhesion, activation, aggregation
Degradation profile, degradation product form, and toxicity (for polymeric materials)	Complete activation
Leachables, additives, catalysts, contaminants, and their toxicity (for polymeric materials)	Antibody production, immune cell response
Dissolution/degradation profile, degradation product toxicity (for ceramic materials)	Acute hypersensibility/anaphylaxis
Wear debris release profile	Delayed hypersensitivity
	Mutagenic response, genotoxicity
	Reproductive toxicity
	Tumor formation

Table 5. Biomaterial variables that could influence the host response [80].

and subcellular levels such as cell proliferation or death in contact with materials [80]. These protocols and methods of interpretation may be used to enhance the information given in the National and International standards.

4.2.1.1. Cell culture testing

The most common and initial evaluation of a new material is by placing the material or an extract of the material into a suitable laboratory cell culture and by observing any changes in the cells over a period of hours to a few days [81]. These tests are performed on primary cell cultures or established cell lines (commercially available), which allows comparison of testing performed for different materials using nearly identical cloned cells. The use of PMMA acrylic base denture has been widely investigated in culture cells alone and enriched with different materials. The enlisted publications in **Table 6** was searched at MEDLINE/PUBMED with the following keywords: "Cytotoxicity AND acrylic resins," "cytotoxicity AND denture

Author	PMMA modification	Culture cells	Assays	Culture time	Results
Herman et al. [82]	PMMA, monomer modified with DABCO (DC16, DC16F, DC18, C6DC16) and conjugated monomers (DC11MAF and C2DC11MAF) at 1, 2, or 3%	Periodontal ligament cells (PDL) and gingival fibroblast (HGF)	BioTek Synergy2 fluorescent	24 h	DABCO components exhibit intermediate to high cytotoxicity and DC11MAF exhibited the lowest toxicity against PDL and GF
Song et al. [83]	PMMA with chitosan (0.50, 1, 2, 3 mg/ml)	Mouse fibroblast cells (L929)	MTT	0, 24, 48, 72 h	No significant difference in cell proliferation between conventional resin and the chitosan quaternary ammonium salt modified
Liu et al. [84]	PMMA-PEI (polyethyleneimine) nanoparticles	Kupffer cells (KCs) primary culture	Cell Counting Kit-8 assay Fluorescence Western blot	6, 12, 24, 48 h	Exhibit survival fraction higher than 90%. These results suggested that the PEI-PMMA/miRNA-complexed NPs had low cytotoxicity to KCs
da Silva et al. [85]	N1 acrylic (MMA polymer, dibutyl phthalate), ethyl acrylate pigments), Poli-Côr (color R2, MMA polymer, dibutyl phthalate, ethyl acrylate, around 1.5% of various organic and inorganic pigments), Clássico (MMA monomer, topanol)	Human conjunctiva cell line (CCL-20.2).	MTT ELISA RT-PCR	72 h	Non-cytotoxic based on cell proliferation. Resin with pigment showed significant increase of IL6
Carlsson et al. [86]	PMMA-based bone cement-Osteopal V modified with castor oil and linoleic acid	Human osteoblast-like Saos-2 cells (HPACC)	AlamarBlue assay Fluorescence	24 h	*In vitro* cytotoxicity appeared somewhat affected by the castor oil and linoleic acid additions
Jiao et al. [87]	PMMA enriched with 15% of N-acetyl cysteine (NAC)	Human dental pulp cells	Extracts preparation and MTT	3, 7 days	The addition of NAC remarkably improved biocompatibility of PMMA resin

Author	PMMA modification	Culture cells	Assays	Culture time	Results
Jang et al. [88]	Heat-polymerized acrylic resin (Paladent 20), thermoplastic acrylic resin (Acrytone and Bio Tone)	Human gingival fibroblasts (HGF) primary cell culture	EZ-Cytox Enhanced Cell Viability Assay Kit Cell attachment (FE-SEM)	1, 6, 10 days	The three types of denture base showed low cytotoxicity in cell viability assay Thermoplastic acrylic resin showed the similar cell attachment but more stable attachment than conventional heat-polymerized acrylic resin
Sahin et al. [89]	PMMA enriched with 2-hydroxyethyl methacrylate (HEMA) and isobutyl methacrylate (IBMA) at 2, 3, and 5%	Mouse fibroblast cells (L929)	Agar overlay test	24 h	Only IBMA showed no cytotoxic effect at low concentrations while HEMA showed cytotoxic effect in the injection-molded resins
Yu [90]	PMMA enriched with calcium phosphate cement (CPC) at 3:1, 2:1, 1:1, 1:2, 1:5, 1:10, 1:15, and 1:20	Osteoblastic progenitor cells (MC3T3-E1)	CCK8 detection reagent	24 h	Bone cement extracts had no effect on the relative MC3T3-E1 cell growth rate, and the toxic reaction was level 1 (75–99%)
Retamoso et al. [91]	Self-curing acrylic resin of different colors: clear, pink, blue, and green	Mouse fibroblast cells (L929)	Dye-uptake technique	24, 48, 72, 168 h	Supernatants evaluation of the color of resin proved not to influence material cytotoxicity
Brochu et al. [92]	PMMA cement (Palcos R bone cement) enriched with microencapsulated 2-octyl cyanoacrylate (OCA); extracts solutions	Human osteosarcoma cells (MG63)	Click-iTEdUAlexa Fluor 488 kit for fluorescence	24, 48, 72 h	Cell proliferation and viability were not significantly different from each other, whereas extracts from OCA were moderately toxic to cells
dos Santos et al. [93]	Acrylic resin (OrtoCril) chemical and mechanical polishing and without polishing	Mouse fibroblast cells (L929)	Dye uptake	24, 48, 72, 168 h	With the increase of cell viability, from the 72 h, there was no significant difference among the groups

Author	PMMA modification	Culture cells	Assays	Culture time	Results
Son et al. [94]	Scaffolds were fabricated by electrospinning using polycaprolactone (PCL) blended with PMMA; extracts solutions	Human osteosarcoma cells (MG63)	MTT Western blot	1, 5, 7 days	PCL/PMMA blends are suitable for osteoblast cell proliferation
Jiang et al. [95]	PMMA particles	Bone marrow stromal cells	MTT CytoTox 96 ELISA RT-PCR.	1, 3, 5, 7 days	PMMA did not stimulate cell proliferation effect even at low doses (0.63mg/ml) and the particles appeared to exhibit certain cytotoxic effect at high concentration (3 mg/ml)
Neves et al. [96]	Ethanol treatment postpolymerization of PMMA, extracts test from heath treatment and conventional	Human adult dermal fibroblast cells	MTT LDH assay	24 h	Specimens showed significant reduction on cytotoxicity compared to immersion in hot water
Tencomnao et al. [97]	PMMA core/polyethyleneimine (PEI) shell magnetic nanoparticles	Human neuroblastoma (LAN-5)	MTT	24 h	The viability of LAN-5 cells after transfection was in the range of 80–100%
Acosta-Torres et al. [98]	PMMA enriched with Silver nanoparticles (AgNPs)	NIH-3T3 mouse embryonic cells	MTT BrdU assay	24, 72 h	Non-cytotoxic material
Tay et al. [99]	Lucitone 550-HR Soft-Liners: Ufi-Gel P-Silicon Dentuflux-AR Trusoft-AR Dentusoft-Tissue conditioner Water storage time after polymerization	Mouse fibroblast cells (L929)	H-thymidine incorporation assay	24 h	Trusoft, lucitone 550 showed slightly cytotoxic effect Dentuflux showed moderate cytotoxic effect when materials were stored in water non-cytotoxic effect Thermal treatment did not reduce the cytotoxicity effect of the acrylic-based soft lines

Author	PMMA modification	Culture cells	Assays	Culture time	Results
Ebrahimi Saravi et al. [100]	AR-Futura Gen AR-GC Reline Hard HR-Meliodent	Mouse fibroblast cells (L929)	MTT ELISA	1 h, 24 h, 1 week	Cytotoxicity of Futura Gen, GC Reline Hard and Meliodent resins failed to show any significant reduction from 24 h to one week. The lower the incubation periods, the higher the cytotoxicity
Regis et al. [101]	MMA MUPB (monomer methacryloyloxyundecylpyridinium bromide)	Mouse fibroblast cells (L929)	MTT	48 h	High concentration of MMA (1g/L) reduces cell viability. MUPB exhibit less cytotoxicity than MMA
Trubiani et al. [102]	Tokuyama Rebase Fast II-AuR IvoclarProbase Cold-AuR Coldpack Tooth Acrylic-AuR polished or unpolished	Human gingival fibroblast (HGF)	MTT ELISA Western blot	24, 48, 72 h	Polishing procedure can reduce the cytotoxicity
Cochis et al. [103]	Paladon 65-HR precoated with biosurfactant	Mouse fibroblast cells (L929), human keratinocytes	MTT	–	Surfactants on resin for prosthetic devices were non-cytotoxic

Table 6. Cytotoxicity test of PMMA alone or modified.

base resins," and "cytotoxicity AND oral prosthesis." Inclusion criteria were: *in vitro* studies published from 2012 to 2017, free full text, and published in English evaluating the PMMA and its components, considering cytotoxicity activity, type of material tested, kinds of cells used, period of incubation, assay executed, and results of the biocompatibility. Two reviewers read the selected studies, and their information was analyzed and discussed. **Figure 5** shows the flow chart of search strategy and the total number of studies included.

4.2.1.2. Mutagenicity testing

A concern for any material used in medicine is that long-term exposure to a material can lead to neoplastic changes in cells adjacent to it. Most materials are known to be acceptable based on a history of use, but changes in formulations and innovation of new materials are necessary to re-execute testing. Small animal *in vivo* mutagenicity studies allow screening of materials in development and reduce the use of research animals. Mutagenicity studies (also called genotoxicity studies) involve looking for changes in cells and cellular DNA in the forward or reverse directions. In forward mutation studies, normal cells are exposed to the test material and the resultant cells or animal tissues are evaluated for signs of mutation. Just a few studies have been tested for genotoxicity between PMMA and culture or small number of animal evaluations has been conducted. **Table 7** summarizes the investigations performed between mutagenicity and PMMA. The keywords used for search strategy at MEDLINE/PUBMED were as follows: "Genotoxicity AND acrylic resin," "genotoxicity AND polymethylmethacrylate resin,"

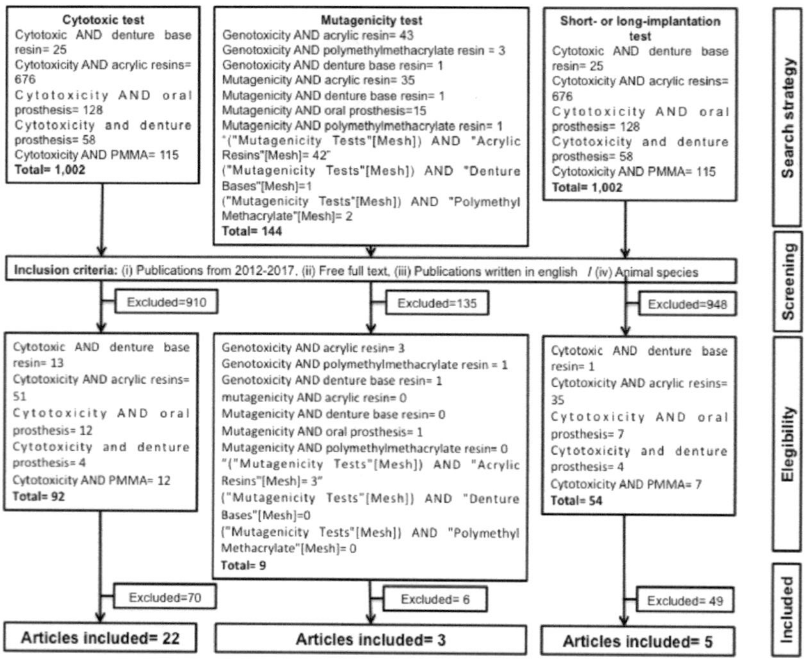

Figure 5. Search strategy flow chart. Source: Direct.

Author	PMMA modification	Culture cells or animal test	Assays	Culture time	Results
Azhar et al. [104]	Methyl methacrylate (MMA) detected in dental lab technicians	Buccal mucosa scrapes (epithelial cells)	Papanicolaou staining Buccal Micronucleus Cytome (BMCyt) assay	Exposure to MMA time of dental lab technicians during their professional career	No significant differences in the incidence of dental lab technicians and control group
Araújo et al. [105]	Methyl methacrylate (MMA) vapor by simulating standard occupational exposure of 8 hours per day	Male Wistar rats	Stained with Giemsa staining (Micronucleus test)	1–5 days	MMA was genotoxic when measured after 1 day of exposure but was not evidently genotoxic after 5 days
Acosta-Torres et al. [98]	PMMA enriched with Silver nanoparticles (AgNPs)	NIH-3T3 mouse embryonic cells	MTT BrdU assay Comet assay	24, 72 h	Non-cytotoxic material

Table 7. Mutagenicity test evaluation of PMMA alone and modified with different materials.

"genotoxicity AND denture base resin," "mutagenicity AND acrylic resin," "mutagenicity AND denture base resin," "mutagenicity AND oral prosthesis," and "mutagenicity AND polymethylmethacrylate resin." The inclusion criteria were: publications from 2012 to 2017, free full text. Only a few studies are reported in literature, a further MeSH term search was executed "("Mutagenicity Tests"[Mesh]) AND "Acrylic Resins"[Mesh], ("Mutagenicity Tests"[Mesh]) AND "Denture Bases"[Mesh], and ("Mutagenicity Tests"[Mesh]) AND "Polymethyl Methacrylate"[Mesh]. **Figure 1** shows the flow chart of search strategy and the total number of studies included.

4.2.1.3. Short- or long-term injection or implantation studies

Several different tests may be conducted to provide information on the effects of relatively short-term exposure to materials or their extracts. These tests include systemic injection, intracutaneous injection for irritation, and short- and long-term implant studies ranging from 24 hours to as long as 90 days or years [80]. Certain materials have the potential to cause local inflammation of tissues. A special case of long-term implantation studies involves the lifetime bioassay performed for investigation of carcinogenicity. These studies are usually performed in several hundred rats and mice to look for differences in tumor formation as a result of exposure to the test material [79]. These studies allow screening out of candidate materials that may not be suitable for further testing. **Table 8** shows the results of search strategy at MEDLINE/PUBMED. The search strategy was previously described in cells tested with the incisive criteria of animal test. **Figure 5** shows the flow chart of search strategy and the total number of studies included.

Author	Type of material	Animal model	Implantation time	Analysis	Results
Carlsson et al. [86]	PMMA-based bone cement-Osteopal V modified with castor oil and linoleic acid	Male Sprague-Dawley rats	1, 4, 12 weeks	Flow cytometry Histological analysis	No differences could be found in the *in vivo* response to these PMMA-based cements
Tsuji et al. [106]	Coating with titanium dioxide (TiO_2) nanoparticles	Hamster oral mucosa irritation test, a guinea pig skin sensitization test and a rabbit intracutaneous test	Irritation test: 24 h Skin sensitization: 2 days Intracutaneous test: 24, 48, and 72 h	Histological analysis	The PMMA coated with TiO2 NPs does not cause irritation or sensitization of the oral mucosa, skin, or intracutaneous tissue and is therefore good
Liu et al. [84]	PMMA-PEI nanoparticles	C57/BL6 mice	6 hours	Western blot NF-κB P65 protein levels in liver tissue	PMMA-PEI NPs could induce targeted transfection (34.7%)
Yu [90]	PMMA enriched with calcium phosphate cement (CPC) at 3:1, 2:1, 1:1, 1:2, 1:5, 1:10, 1:15, and 1:20	SD rats bone defect	4 weeks, 15 weeks	X-ray Histological observation	Except for the PMMA group significant degradations appeared in both the CPC/PMMA group (50%; 1:1) and CPC group. Enhanced the bone cell growth
Son et al. [94]	Scaffolds were fabricated by electrospinning using polycaprolactone (PCL) blended with poly(methylmethacrylate) (PMMA)	Sprague Dawley rats; skull defects and PCL/PMMA implantation	1 and 2 months	Micro CT Histological observation	Bone formation was observed on the 7/3 PCL/PMMA scaffold within 2 months

Table 8. Short- or long-term exposure test of PMMA alone or modified.

4.2.2. *Acrylic resin cytotoxicity*

Different methods are used for cytotoxicity, mutagenicity, and short- or long-term implantation analysis in the literature. Among them, the most common is the MTT test [3-(4,5-dimethylthiazol-2-yl)-2,5-diphenyltetrazolium] and a histological evaluation. In the case of MTT, the method quantifies the mitochondrial succinate dehydrogenase enzyme activity and measures the conversion of water-soluble tetrazolium salt in insoluble blue formazan by spectrophotometry. This test is an excellent marker of cell survival because it evaluates cellular respiratory activity [100, 107].

The cytotoxicity of PMMA is correlated with the polymerization methods, temperature, the cycle of polymerization, and acrylic resin storage time can influence the monomer quantity and the material cytotoxicity [95, 100]. Based on the polymerization method, acrylic resin can be classified as heat-polymerized, microwave-polymerized, light-polymerized, and autopolymerized MMA. The latter being the most commonly used in dental practice [15]. The autopolymerized resin exhibited higher cytotoxicity level than heat-polymerized resin after 1 and 24 h of incubation [100]. The experiment performed of PMMA alone in contact with HGF showed similar results of dose-dependent cytotoxicity. It is important to use polished acrylic resins for clinical applications. The unpolished acrylic resin showed cell growth reduction, and an increase in pro-inflammatory cytokines were caused by the tested material [93, 102].

Postpolymerization heat treatments, such as water bath or microwave irradiation, have been suggested in order to reduce the quantity of autopolymerized acrylic resin residual monomers. The PMMA that was immersed into water showed a reduction in MMA monomer elucidation [88, 96].

Several substances such as chitosan [83], PEI (polyethyleneimine) nanoparticles [84], 15% of N-acetyl cysteine (NAC) [87], calcium phosphate cement (CPC) [90], acrylic resin of different colors [91], (bone cement) enriched with microencapsulated 2-octyl cyanoacrylate (OCA), extracts solutions [92], scaffolds fabricated by electrospinning using polycaprolactone (PCL) [94], core/polyethyleneimine (PEI) shell magnetic nanoparticles [97], silver nanoparticles (AgNPs) [98], and Paladon 65-HR precoated with biosurfactant were also evaluated for cytotoxicity. The authors observed reduction in cytotoxicity and increase in biocompatibility from non-cytotoxic (cell viability higher than 75%) to slightly cytotoxic (cell viability ranging from 50 to 75%)

On the other hand, DABCO (DC16, DC16F, DC18, C6DC16) and conjugated monomers (DC11MAF and C2DC11MAF) at 1, 2, or 3% [82], base bone cement-Osteopal V modified with castor oil and linoleic acid [86], 2-hydroxyethyl methacrylate (HEMA) and isobutyl methacrylate (IBMA) at 2, 3, and 5% [89], and MUPB (monomer methacryloyloxyundecylpyridinium bromide) [101] showed cytotoxic effect from moderately cytotoxic (cell viability ranging from 25 to 50%) to severely cytotoxic (cell viability lower than 25%).

Only three studies were reported about the genotoxicity test where the exposure of occupation time did not show difference from patients without continuous exposure or cell culture [98, 105]. It is necessary to include genotoxicity assay for further investigations of potential biomaterials in dental practice.

Several studies have been carried out at short- or long- term implantation with PMMA enriched or coated with different materials as base bone cement-osteopal V modified with castor oil and linoleic acid in rats [86], coating with titanium dioxide (TiO_2) nanoparticles for hamster oral mucosa irritation and guinea pig skin sensitization and intracutaneous rabbit implantation [106], enriched with CPC in rats [90], scaffolds of electrospinning using polycaprolactone (PCL) implantation in rats [94], no toxic effect or histological findings were observed, even a regeneration was perceived. By contrast, the use of PMMA-PEI nanoparticles injected in mice induces significant toxicity by the detection of protein levels in liver tissue.

Acrylic resin cytotoxicity is associated with the presence of residual monomer in the polymerization process. The monomers change cell morphology and function that can reduce their viability. Since acrylic resins are widely used in dental practice, an acceptable biocompatibility is desirable. Considering that the majority of studies reported acrylic resin toxicity responses, further studies with different assessment methods are necessary for the development of biocompatible materials.

In summary, there exist different methods to evaluate acrylic resin cytotoxicity, genotoxicity, and short- or long-term implantation with the MTT method and histological evaluation being the most common tests. In conclusion, there is no non-cytotoxic acrylic resin evidently available in the dental market. Regarding the methods of polymerization, the autopolymerized resin is more cytotoxic and toxic than heat-polymerized resin. The cytotoxic and toxicity effect is dose dependent and is directly correlated with the residual amount of monomer leachable and induce the inflammatory reactions of tissues in contact with the acrylic resin. It is suggested that a water or ethanol bath after polymerization of acrylic resin could decrease the cytotoxic and toxicity activity against oral cells and tissue.

5. Remarks and perspectives

The particle size differences could influence the roughness surface of PMMA without decreasing their mechanical properties. Furthermore, studies should continue to determine that issue in detail.

Ultrasound can be used to process the acrylic resin (Opticryl®) as an alternative technique for PMMA processing with similar results to those obtained using water bath or microwave processing (control groups).

The addition of antifungal agents on PMMA surface or into PMMA, such as moieties, metallic or metallic oxide nanoparticles, and antibiotics, could be useful for the inhibition of specific pathogens such as *Candida albicans* for prosthodontic denture.

Residual monomer leach induces cytotoxic and inflammatory reactions of oral cells and tissues in contact with the acrylic resin. It is suggested that water or ethanol bath after polymerization of acrylic resin could decrease the cytotoxic and toxicity activity against oral cells and tissues. But ultrasonic waves are a good option for both, for thermopolymerization of PMMA and at the same time to reduce the residual monomer.

Acknowledgements

The authors want to thank PAPIME-DGAPA, UNAM for its financial support through projects PE205916, PE210616, and PAPIIT-DGAPA (IN225526, IA204516). Thanks to José Campos (IER-UNAM), Ana Cecilia Pérez, Neftalí Hernández, Carlos García, and Ma. Elena Calixto (ITESI) for excellent technical support. L. Argueta-Figueroa thanks DGAPA, UNAM for the Postdoctoral scholarship.

Author details

Ma. Concepción Arenas-Arrocena*, Liliana Argueta-Figueroa, René García-Contreras, Omar Martínez-Arenas, Berenice Camacho-Flores, María del Pilar Rodriguez-Torres, Javierde la Fuente-Hernández and Laura S. Acosta-Torres

*Address all correspondence to: carenas@enes.unam.mx

National School of Higher Education Unit Leon, National Autonomous University of Mexico, Guanajuato, Mexico

References

[1] Thomason JM, Heydecke G, Feine JS, Ellis JS. How do patients perceive the benefit of reconstructive dentistry with regard to oral health-related quality of life and patient satisfaction? A systematic review. Clinical Oral Implants Research. 2007;**18**(3):168-188. DOI: 10.1111/j.1600-0501.2007.01461.x

[2] Müller F, Salem K, Barbezat C, Schimmel M. Knowledge and attitude of elderly persons towards dental implants. Gerodontology. 2012;**29**:e914-e923. DOI: 10.1111/j.1741-2358.2011.00586.x

[3] Kim Y, Park JY, Park SY, Oh SH, Jung Y, Kim JM, Yoo SY, Kim SK. Economic evaluation of single-tooth replacement: Dental implant versus fixed partial denture. International Journal of Oral & Maxillofacial Implants. 2014;**29**:600-607. DOI: 10.11607/jomi.3413

[4] McKenna G, Allen PF, O'Mahony D, Cronin M, DaMata C, Woods N. The impact of rehabilitation using removable partial dentures and functionally orientated treatment on oral health-related quality of life: A randomised controlled clinical trial. Journal of Dentistry. 2015;**43**:66-71. DOI: 10.1016/j.jdent.2014.06.006

[5] Tandon R, Gupta S, Agarwal S. Denture base materials: From past to future. Indian Journal of Dental Sciences. 2010;**2**:33-39

[6] Peyton FA. History of resins in dentistry. Dental Clinic of North America. 1975;**19**(2):211-222

[7] Takahashi N. Oral microbiome metabolism: From "who are they?" to "what are they doing?". Journal of Dental Research. 2015;**94**:1628-1637. DOI: 10.1177/0022034515606045

[8] Gendreau L, Loewy ZG. Epidemiology and etiology of denture stomatitis. Journal of Prosthodontics. 2011;**20**:251-260. DOI: 10.1111/j.1532-849X.2011.00698.x

[9] Do Amaral BA, Barreto AO, Gomes-Seabra E, Roncalli AG, Porte-Carreiro AF, Olveira-de Almeida E. A clinical follow-up study of the periodontal conditions of RPD abutment and non-abutment teeth. Journal of Oral Rehabilitation. 2010;**37**:545-552. DOI: 10.1111/j.1365-2842.2010.02069.x

[10] Augustin MM, Joke D, Bourleyi SI, Shenda LP, Fidele NB, Gabriel BB, Pierre SN, Kazadi EK, Ediz EI, Pierrot KN, Mbuebo M, Kanda NM, Naert I. Risks factors of caries and periodontal diseases in the patients, after 5 years use a partial removable denture keywords partial removable denture acrylic resin, caries and periodontal diseases, oral health related quality of life. Open Journal of Stomatology. 2016;**6**(8):185-192. DOI:10.4236/ojst.2016.68024

[11] Bhola R, Bhola SM, Liang H, Mishra B. Biocompatible denture polymers—A review. Trends in Biomaterials and Artificial Organs. 2010;**23**(3):129-136

[12] Ayaz EA, Durkan R, Koroglu A, Bagis B. Comparative effect of different polymerization techniques on residual monomer and hardness properties of PMMA-based denture resins. Journal of Applied Biomaterials & Functional Materials. 2014;**12**:228-233. DOI: 10.5301/jabfm.5000199

[13] Wang J, Chen C, Buck SM, Chen Z. Molecular chemical structure on poly (methyl methacrylate) (PMMA) surface studied by sum frequency generation (SFG) vibrational spectroscopy. The Journal of Physical Chemistry B. 2001;**105**(48):12118-12125. DOI: 10.1021/jp013161d

[14] Frazer RQ, Byron RT, Osborne PB, West KP. PMMA: An essential material in medicine and dentistry. Journal of Long-Term Effects of Medical Implants. 2005;**15**:629-639

[15] Gautam R, Singh RD, Sharma VP, Siddhartha R, Chand P, Kumar R. Biocompatibility of polymethylmethacrylate resins used in dentistry. Journal of Biomedical Materials Research Part B, Applied Biomaterials. 2012;**100**(5):1440-1450. DOI: 10.1002/jbm.b.32673

[16] Chaiyasat P, Namwong S, Okubo M, Chaiyasata A. Synthesis of micrometer-sized poly(methyl methacrylate) particles by microsuspension iodine transfer polymerization (ms ITP). RSC Advances. 2016;**6**:95062-95066. DOI: 10.1039/C6RA19288B

[17] Albrecht K, Manfred Strickler TR. Ullman's Encyclopedia of Industrial Chemistry, Weinheim, Germany. Wiley-VCH Verlag GmbH & Co. KGaA; 2013. DOI: 10.1002/14356007.a21_473.pub2.

[18] Lyoo WS, Noh SK, Yeum JH, Kang GC, Ghim HD, Lee J, Ji BC. Preparation of high molecular weight poly(methyl methacrylate) with high yield by room temperature suspension polymerization of methyl methacrylate. Fibers and Polymers. 2004;**5**(75):75-81. DOI: 10.1007/BF02875498.

[19] Grimaud T, Matyjaszewski K. Controlled/"Living" radical polymerization of methyl methacrylate by atom transfer radical polymerization. Macromolecules. 1997;**30:**2216-2218. DOI: 10.1021/ma961796i

[20] Odian G. Principles of polymerization. 4th ed. New York: John Wiley & Sons, Inc., College of Staten Island, City University of New York Staten Island; 2004

[21] Harry L, Frederick WA, Lampe JEM. Contemporary Polymer Chemistry. 2nd ed. Englewood Cliffs: Prentice Hall; 1981. DOI: 10.1002/pol.1981.130190509

[22] Dowding PJ, Vincent B. Suspension polymerisation to form polymer beads. Colloids and Surfaces A: Physicochemical and Engineering Aspects. 2000;**161**:259-269. DOI: 10.1016/S0927-7757(99)00375-1

[23] Hoffman F, Delbruch K. Patent (Ger.) No. 250 690, Farbenfabriken Bayer; 1909

[24] Moreno-Maldonado V, Acosta-Torres LS, Barceló-Santana FH, Vanegas-Lancón RD, Plata-Rodríguez ME, Castaño VM. Fiber-reinforced nanopigmented poly(methyl methacrylate) as improved denture base. Journal of Applied Polymer Science. 2014;**132**(2):289-296. DOI: 10.1002/app.36913

[25] Acosta -Torres LS, López-Marín LM, Núñez-Anita RE, Hernández-Padrón G, Castaño VM. Biocompatible metal-oxide nanoparticles: Nanotechnology improvement of conventional prosthetic acrylic resins. Journal of Nanomaterials. 2011;ID 941561, 8 pages. DOI: 10.1155/2011/941561

[26] Acosta-Torres LS, Arenas MC, Nuñez-Anita RE, Barceló-Santana FH, Álvarez-Gayosso CA, Palacios-Alquisira J, De La Fuente-Hernández J, Cajero-Juárez M, Castaño VM. Nanopigmented acrylic resin cured indistinctively by water bath or microwave energy for dentures. Journal of Nanomaterials. 2014;ID 198572, 8 pages. DOI: 10.1155/2014/198572

[27] Baissac L, Buron CC, Hallez L, Berçot P, Hihn JY, Chantegrel L, Gosse G. Synthesis of sub-micronic and nanometric PMMA particles via emulsion polymerization assisted by ultrasound: Process flow sheet and characterization. Ultrasonics Sonochemistry. 2017. in press. DOI: 10.1016/j.ultsonch.2017.03.027

[28] Senawongse P, Pongprueksa P. Surface roughness of nanofill and nanohybrid resin composites after polishing and brushing. Journal Compilation. 2007;**19**(5):265-273. DOI: 10.1111/j.1708-8240.2007.00116.x

[29] Ali U, Karim KJ. Bt, Buang NA. A review of the properties and applications of poly(methyl methacrylate) (PMMA). Polymer Reviews. 2015;**55**(4):678-705. DOI: 10.1080/15583724.2015.1031377.

[30] Pero AC, Barbosa DB, Marra J, Ruvolo-Filho AD, Compagnoni MA. Influence of microwave polymerization method and thickness on porosity of acrylic resin. Journal of Prosthodontics. 2008;**17**:125-129. DOI: 10.1111/j.1532-849X.2007.00264.x

[31] Kasina SP, Ajaz T, Attili S, Surapaneni H, Cherukuri M, Srinath HP. To evaluate and compare the porosities in the acrylic mandibular denture bases processed by two different

polymerization techniques, using two different brands of commercially available denture base resins—An in vitro study. Journal of International Oral Health. 2014;**6**(1):72-77. PMCID: PMC3959141

[32] Lung CYK, Darvell BW. Minimization of the inevitable residual monomer in denture base acrylic. Dental Materials. 2005;**21**:1119-1128. DOI: 10.1016/j.dental.2005.03.003

[33] Ganzarolli SM, Nunes de Mello JA, Shinkai RS, Del Bel Cury AA. Internal adaptation and some physical properties of methacrylate-based denture base resins polymerized by different techniques. Journal of Biomedical Materials Research Part B Applied Biomaterials. 2007;**82B**:169-173. DOI: 10.1002/jbm.b.30718

[34] Jadhav R, Bhide SV, Prabhudesai PS. Assessment of the impact strength of the denture base resin polymerized by various processing techniques. Indian Journal of Dental Research. 2013;**24**:19-25. DOI: 10.4103/0970-9290.114926

[35] Banerjee R, Banerjee S, Prabhudesai P, Bhide SV. Influence of the processing technique on the flexural fatigue strength of denture base resins: An in vitro investigation. Indian Journal of Dental Research. 2010;**21**(3):391-395. DOI: 10.4103/0970-9290.70810

[36] Francis J, Shetty R, Shenoy K, Somaraj V. Comparative evaluation on the influence of different curing cycles on the mechanical properties of three commercially available denture base resins: An in-vitro study. Journal of Applied Dental and Medical Sciences. 2016;**3**:23-30

[37] Çelebi N, Yüzügüllü B, Canay Ş, Yücel Ü. Effect of polymerization methods on the residual monomer level of acrylic resin denture base polymers. Polymers for Advanced Technologies. 2008;**19**:201-206. DOI: 10.1002/pat.996

[38] De Clerck JP. Microwave polymerization of acrylic resins used in dental prostheses. The Journal of Prosthetic Dentistry. 1987;**57**:650-658. PMID: 3298632

[39] Lai CP, Tsai MH, Chen M, Chang HS, Tay HH. Morphology and properties of denture acrylic resins cured by microwave energy and conventional water bath. Dental Materials. 2004;**20**:133-141. PubMed: 14706796

[40] Jacob J, Chia LHL, Boey FYC. Comparative study of methyl methacrylate cure by microwave radiation versus thermal energy. Polymer Testing. 1995;**14**(4):343-354. DOI: doi.org/10.1016/0142-9418(94)00034-C

[41] Spartalis GK, Cappelletti LK, Schoeffel AC, Michél MD, Pegoraro TA, Arrais CA, Neppelenbroek KH, Urban VM. Effect of conventional water-bath and experimental microwave polymerization cycles on the flexural properties of denture base acrylic resins. Dental Materials Journal. 2015;**34**:623-628. DOI: 10.4012/dmj.2015-047

[42] Hashem M, Alsaleem SO, Assery MK, Abdeslam EB, Vellappally S, Anil S. A comparative study of the mechanical properties of the light-cure and conventional denture base resins. Oral Health and Dental Management. 2014;**13**:311-315. PMID: 24984639

[43] Alkhatib MB, Goodacre CJ, Swartz ML, Munoz-Viveros CA, Andres CJ. Comparison of microwave-polymerized denture base resins. International Journal of Prosthodontics. 1990;**3**:249-255. PMID: 2083011

[44] Archadian N, Kawano F, Ohguri T, Ichikawa T, Matsumoto N. Flexural strength of rebased denture polymers. Journal of Oral Rehabilitation. 2000;**27**:690-696. PubMed: 10931264

[45] Smith LT, Powers JM, Ladd D. Mechanical properties of new denture resins polymerized by visible light, heat, and microwave energy. The International Journal of Prosthodontics. 1992;**5**:315-20. PMID: 1520453

[46] Barbosa DB, Souza RF, Pero AC, Marra J, Compagnoni MA. Flexural strength of acrylic resins polymerized by different cycles. Journal of Applied Oral Science. 2007;**15**:424-428. DOI: 10.1590/S1678-77572007000500010

[47] Compagnoni MA, Barbosa DB, Souza RF, Pero AC. The effect of polymerization cycles on porosity of microwave-processed denture base resin. The Journal of Prosthetic Dentistry. 2004;**91**:281-285. DOI: 10.1016/S0022391304000083

[48] Bafile M, Graser GN, Myers ML, Li EK. Porosity of denture resin cured by microwave energy. The Journal of Prosthetic Dentistry. 1991;**66**:269-274. PMID: 1774691

[49] Charasseangpaisarn T, Wiwatwarrapan C, Leklerssiriwong N. Ultrasonic cleaning reduces the residual monomer in acrylic resins. Journal of Dental Sciences. 2016;**11**:443-448. DOI: dx.doi.org/10.1016/j.jds.2016.07.003

[50] Henglein A. Sonochemistry: Historical developments and modern aspects. Ultrasonics. 1987;**25**:6-16. DOI: 10.1016/0041-624X(87)90003-5

[51] Bang JH, Suslick KS. Applications of ultrasound to the synthesis of nanostructured materials. Advanced Materials. 2010;**22**:1039-1059. DOI: 10.1002/adma.200904093

[52] Suslick KS, Flannigan DJ. Inside a collapsing bubble: Sonoluminescence and the conditions during cavitation. Annual Review of Physical Chemistry. 2008;**59**:659-683. DOI: 10.1146/annurev.physchem.59.032607.093739

[53] Bhanvase BA, Sonawane SH. Ultrasound assisted in situ emulsion polymerization for polymer nanocomposite: A review. Chemical Engineering and Processing. 2014;**85**:86-107. DOI: dx.doi.org/10.1016/j.cep.2014.08.007

[54] Charasseangpaisarn T, Wiwatwarrapan C. The effect of various frequencies of ultrasonic cleaner in reducing residual monomer in acrylic resin. Ultrasonics. 2015;**63**:163-167. DOI: 10.1016/j.ultras.2015.07.005

[55] Oliver WC, Pharr GM. An improved technique for determining hardness and elastic modulus using load and displacement sensing indentation experiments. Journal of Materials Research. 1992;**7**:1564-1583. DOI: doi.org/10.1557/JMR.1992.1564

[56] Chitchumnong P, Brooks SC, Stafford GD. Comparison of three- and four-point flexural strength testing of denture-base polymers. Dental Materials. 1989;**5**:2-5. PMID: 2606263

[57] Allaker RP, Memarzadeh K. Nanoparticles and the control of oral infections. International Journal of Antimicrobial Agents. 2014;**43**(2):95-104. DOI: doi.org/10.1016/j.ijantimicag.2013.11.002

[58] Ivanova EP, Hasan J, Webb HK, Gervinskas G, Juodkazis S, Truong VK, Wu AHF, Lamb RN, Baulin VA, Watson GS, Watson JA, Mainwaring DE, Crawford RJ. Bactericidal activity of black silicon. Nature Communications. 2013;**4**:2838. DOI: doi.org/10.1038/ncomms3838

[59] Hasan J, Crawford RJ, Ivanova EP. Antibacterial surfaces: The quest for a new generation of biomaterials. Trends in Biotechnology. 2013;**31**(5):295-304. DOI: doi.org/10.1016/j.tibtech.2013.01.017

[60] Ivanova EP, Truong VK, Wang JY, Bemdt CC, Jones RT, Yusuf II, Peake I, Schmidt HW, Fluke C, Barnes D, Crawford RJ. Impact of nanoscale roughness of titanium thin film surfaces on bacterial Retention. Langmuir. 2010;**26**(3):1973-1982. DOI: doi.org/10.1021/la902623c

[61] Mandal BK. Nanobiomaterials in antimicrobial therapy. Applications of Nanobiomaterials. 2016;**6**: 313-341. DOI: doi.org/10.1016/B978-0-323-42864-4.00009-9

[62] Cierech M, Kolenda A, Grudniak AM, Wojnarowicz J, Woźniak B, Gołaś M, Swoboda-Kopeć E, Łojkowski W, Mierzwińska-Nastalska E. Significance of polymethylmethacrylate (PMMA) modification by zinc oxide nanoparticles for fungal biofilm formation. International Journal of Pharmaceutics. 2016;**510**:323-335. DOI: 10.1016/j.ijpharm.2016.06.052

[63] Wang W, Liao S, Zhu Y, Liu M, Zhao Q, Fu Y. Recent applications of nanomaterials in prosthodontics. Journal of Nanomaterials. 2015;ID 408643, 11 pages. DOI: 10.1155/2015/408643.

[64] Lemire JA, Harrison JJ, Turner RJ. Antimicrobial activity of metals: Mechanisms, molecular targets and applications. Nature Reviews. Microbiology. 2013;**11**(6):371-84. DOI: doi.org/10.1038/nrmicro3028

[65] Sheth SR, Leckband D. Measurements of attractive forces between proteins and end-grafted poly(ethylene glycol) chains. Proceedings of the National Academy of Sciences of the United States of America. 1997;**94**(16):8399-8404. DOI: doi.org/0027-8424?97?948399-6

[66] Al-Dwairi ZN, Al-Quran FA, Al-Omari OY. The effect of antifungal agents on surface properties of poly(methyl methacrylate) and its relation to adherence of *Candida albicans*. Journal of Prosthodontic Research. 2012;**56**:272-280. DOI: 10.1016/j.jpor.2012.02.006

[67] Labarre D. The interactions between blood and polymeric nanoparticles depend on the nature and structure of the hydrogel covering the surface. Polymers. 2012;**4**(2):986-996. DOI: doi.org/10.3390/polym4020986

[68] Franci G, Falanga A, Galdiero S, Palomba L, Rai M, Morelli G, Galdiero M. Silver nanoparticles as potential antibacterial agents. Molecules. 2015;**20**(5):8856-8874. DOI: doi.org/10.3390/molecules20058856

[69] Pasquale T, Tan J. Nonantimicrobial effects of antibacterial agents. Clinical Infectious Diseases. 2005;**40**(1):127-135. DOI: doi.org/10.1097/01.inf.0000159186.41813.c1

[70] Cloutier M, Mantovani D, Rosei F. Antibacterial coatings: Challenges, perspectives, and opportunities. Trends in Biotechnology. 2015;**33**:637-652. DOI: doi.org/10.1016/j.tibtech.2015.09.002

[71] Namba N, Yoshida Y, Nagaoka N, Takashima S, Matsuura-Yoshimoto K, Maeda H, Meerbeek BV, Suzuki K, Takashiba S. Antibacterial effect of bactericide immobilized in resin matrix. Dental Materials. 2009;**25**(4):424-430. DOI: doi.org/10.1016/j.dental.2008.08.012

[72] Harini P, Mohamed K, Padmanabhan TV. Effect of Titanium dioxide nanoparticles on the flexural strength of polymethylmethacrylate: An in vitro study. Indian Journal of Dental Research. 2014;**25**(4):459-463. DOI: 10.4103/0970-9290.142531

[73] Nazirkar G, Bhanushali S, Singh S, Pattanaik B, Raj N. Effect of anatase titanium dioxide nanoparticles on the flexural strength of heat cured polymethylmethacrylate resins: An in vitro study. Journal of Indian Porsthodontic Society. 2014;**14**(1):144-149. DOI: 10.1007/s13191-014-0385-8

[74] Shirkavand S, Moslehifard E. Effect of TiO_2 nanoparticles on tensile strength of dental acrylic resins. Journal of Dental Research, Dental Clinics, Dental Prospects. 2014;**8**(4):197-203. DOI: 10.5681/joddd.2014.036

[75] Totu EE, Nechifor AC, Nechifor G, Aboul-Enein HY, Cristache C. Poly (methymethacrylate) with TiO_2 nanoparticles inclusion for stereolithographic complete denture manufacturing—The future in dental care for elderly edentulous patients? Journal of Dentistry. 2017;**59**:68-77. DOI: 10.1016/j.jdent.2017.02.012

[76] Donaruma LG. Definitions in biomaterials, D. F. Williams, Ed., Elsevier, Amsterdam, 1987, 72 pages. Journal of Polymer Science: Polymer Letters Edition. 1988:**26**(9):414-414. DOI: doi.org/10.1002/pol.1988.140260910

[77] Wataha JC. Principles of biocompatibility for dental practitioners. The Journal of Prosthetic Dentistry. 2001;**86**(2):203-209. DOI: doi.org/10.1067/mpr.2001.117056

[78] Porter JA, Von Fraunhofer JA. Success or failure of dental implants? A literature review with treatment considerations. General Dentistry. 2005;**53**(6):423-432. Retrieved from http://www.ncbi.nlm.nih.gov/pubmed/16366052

[79] Van Blitterswijk C, Thomsen P, Lindahl A, Hubbell J, Williams DF, Cancedda R, Williams D. Biocompatibility. Tissue Engineering. 2008;**9**:255-278. DOI: dx.doi.org/10.1016/B978-0-12-370869-4.00009-4

[80] Williams DF. On the mechanisms of biocompatibility. Biomaterials. 2008;**29**(20):2941-2953. DOI: doi.org/10.1016/j.biomaterials.2008.04.023

[81] Schmalz G, Arenholt-Bindslev D. Biocompatibility of Dental Materials. New York: Springer; 2009. DOI: doi.org/10.1007/978-3-540-77782-3

[82] Herman JL, Wang Y, Lilly E, Lallier T, Hamdan S, Xu X, Noverr MC. Synthesis, antifungal activity, and biocompatibility of novel DABCO compounds and DABCO-containing denture base resins. Antimicrobial Agents and Chemotherapy. 2017;**61**:e02575-16. DOI: doi.org/10.1128/AAC.02575-16

[83] Song R, Zhong Z, Lin L. Evaluation of chitosan quaternary ammonium salt-modified resin denture base material. International Journal of Biological Macromolecules. 2016;**85**:102-110. DOI: doi.org/10.1016/j.ijbiomac.2015.12.052

[84] Liu Z, Niu D, Zhang J, Zhang W, Yao Y, Li P, Gong J. Amphiphilic core-shell nanoparticles containing dense polyethyleneimine shells for efficient delivery of microRNA to Kupffer cells. International Journal of Nanomedicine. 2016;**11**:2785-2797. DOI: doi.org/10.2147/IJN.S101251

[85] da Silva EVF, Goiato MC, Bonatto LDR, de Medeiros RA, dos Santos DM, Rangel EC, de Oliveira SHP. Toxicity analysis of ocular prosthesis acrylic resin with or without pigment incorporation in human conjunctival cell line. Toxicology In Vitro. 2016;**36**:180-185. DOI: doi.org/10.1016/j.tiv.2016.08.005

[86] Carlsson E, Mestres G, Treerattrakoon K, López A, Karlsson Ott M, Larsson S, Persson C. In vitro and in vivo response to low-modulus PMMA-based bone cement. BioMed Research International. 2015;ID 594284, 9 pages. DOI: doi.org/10.1155/2015/594284

[87] Jiao Y, Ma S, Li J, Shan L, Yang Y, Li M, Chen J. The influences of N-acetyl cysteine (NAC) on the cytotoxicity and mechanical properties of poly-methylmethacrylate (PMMA)-based dental resin. PeerJ. 2015;**3**:e868. DOI: doi.org/10.7717/peerj.868

[88] Jang D-E, Lee J-Y, Jang H-S, Lee J-J, Son M-K. Color stability, water sorption and cytotoxicity of thermoplastic acrylic resin for non metal clasp denture. The Journal of Advanced Prosthodontics. 2015;**7**(4):278-287. DOI: doi.org/10.4047/jap.2015.7.4.278

[89] Sahin O, Ozdemir AK, Turgut M, Boztug A, Sumer Z. Investigation of flexural strength and cytotoxicity of acrylic resin copolymers by using different polymerization methods. The Journal of Advanced Prosthodontics. 2015;**7**(2):98-107. DOI: doi.org/10.4047/jap.2015.7.2.98

[90] Yu B. Preparation of calcium phosphate cement and polymethyl methacrylate for biological composite bone cements. Medical Science Monitor. 2015;**21**:1162-1172. DOI: doi.org/10.12659/MSM.893845

[91] Retamoso LB, De Morais T, Pithon MM, Lacerda R. In vitro cytotoxicity of self-curing acrylic resins of different colors. Dental Press Journal Orthodontics. 2014;**19**(4):66-70. DOI: doi.org/10.1590/2176-9451.19.4.066-070.oar

[92] Brochu ABW, Evans GA, Reichert WM. Mechanical and cytotoxicity testing of acrylic bone cement embedded with microencapsulated 2-octyl cyanoacrylate. Journal of Biomedical Materials Research Part B Applied Biomaterials. 2014;**102**(1):181-189. DOI: doi.org/10.1002/jbm.b.32994

[93] dos Santos RL, Pithon MM, Carvalho FG, Ramos AADS, Romanos MTV. Mechanical and biological properties of acrylic resins manipulated and polished by different methods. Brazilian Dental Journal. 2013;**24**(5):492-497. DOI: doi.org/10.1590/0103-6440201302293

[94] Son S-R, Linh N-TB, Yang H-M, Lee B-T. In vitro and in vivo evaluation of electrospun PCL/PMMA fibrous scaffolds for bone regeneration. Science and Technology of Advanced Materials. 2013;**14**(1):15009, 10 pages. DOI: doi.org/10.1088/1468-6996/14/1/015009

[95] Jiang Y, Jia T, Gong W, Wooley PH, Yang S-Y. Effects of Ti, PMMA, UHMWPE, and Co-Cr wear particles on differentiation and functions of bone marrow stromal cells. Journal of Biomedical Materials Research Part A. 2013;**101**(10):2817-2825. DOI: doi.org/10.1002/jbm.a.34595

[96] Neves CB, Lopes LP, Ferrão HF, Miranda JP, Castro MF, Bettencourt AF. Ethanol postpolymerization treatment for improving the biocompatibility of acrylic reline resins. BioMed Research International. 2013;ID 485246, 9 pages. DOI: doi.org/10.1155/2013/485246

[97] Tencomnao T, Klangthong K, Pimpha N, Chaleawlert-Umpon S, Saesoo S, Woramongkolchai N, Saengkrit N. Acceleration of gene transfection efficiency in neuroblastoma cells through polyethyleneimine/poly(methyl methacrylate) core-shell magnetic nanoparticles. International Journal of Nanomedicine. 2012;**7**:2783-2792. DOI: doi.org/10.2147/IJN.S32311

[98] Acosta-Torres LS, Mendieta I, Nuñez-Anita RE, Cajero-Juárez M, Castaño VM. Cytocompatible antifungal acrylic resin containing silver nanoparticles for dentures. International Journal of Nanomedicine. 2012;**7**:4777-4786. DOI: doi.org/10.2147/IJN.S32391

[99] Tay LY, Herrera DR, Quishida CCC, Carlos IZ, Jorge JH. Effect of water storage and heat treatment on the cytotoxicity of soft liners. Gerodontology. 2012;**29**(2):275-280. DOI: doi.org/10.1111/j.1741-2358.2011.00463.x

[100] Ebrahimi Saravi M, Vojdani M, Bahrani F. Evaluation of cellular toxicity of three denture base acrylic resins. Journal of Dentistry. 2012;**9**(4):180-188. Retrieved from http://www.ncbi.nlm.nih.gov/pubmed/23323179

[101] Regis RR, Della Vecchia MP, Pizzolitto AC, Compagnoni MA, Souza PPC, de Souza RF. Antimicrobial properties and cytotoxicity of an antimicrobial monomer for application in prosthodontics. Journal of Prosthodontics. 2012;**21**(4):283-290. DOI: doi.org/10.1111/j.1532-849X.2011.00815.x

[102] Trubiani O, Toniato E, Iorio DDI, Diomede F, Merciaro I, Arcangelo CD, Caputi S. Morphological analysis and interleukin release in human gingival fibroblasts seeded on different denture base acrylic resins. International Journal of Immunopathology and Pharmacology. 2012;**25**(3):637-639. http://journals.sagepub.com/doi/pdf/10.1177/039463201202500310

[103] Cochis A, Fracchia L, Martinotti MG, Rimondini L. Biosurfactants prevent in vitro *Candida albicans* biofilm formation on resins and silicon materials for prosthetic devices. Oral Surgery, Oral Medicine, Oral Pathology and Oral Radiology. 2012;**113**(6):755-761. DOI: doi.org/10.1016/j.oooo.2011.11.004

[104] Azhar DA, Syed S, Luqman M, Ali AA. Evaluation of methyl methacrylate monomer cytotoxicity in dental lab technicians using buccal micronucleus cytome assay. Dental Materials Journal. 2013;**32**(3):519-521. http://www.ncbi.nlm.nih.gov/pubmed/23719017

[105] Araújo AM, Alves GR, Avanço GT, Parizi JL, Nai G. Assessment of methyl methacrylate genotoxicity by the micronucleus test. Brazilian Oral Research. 2013;**27**(1):31-36. DOI: doi.org/10.1590/S1806-83242013000100006

[106] Tsuji M, Ueda T, Sawaki K, Kawaguchi M, Sakurai K. Biocompatibility of a titanium dioxide-coating method for denture base acrylic resin. Gerodontology. 2016;**33**(4):539-544. DOI: doi.org/10.1111/ger.12204

[107] Young FM, Phungtamdet W, Sanderson BJS. Modification of MTT assay conditions to examine the cytotoxic effects of amitraz on the human lymphoblastoid cell line, WIL2NS. Toxicology In Vitro. 2005;**19**(8):1051-1059. DOI: doi.org/10.1016/j.tiv.2005.05.001

Chapter 4

Latest Improvements of Acrylic-Based Polymer Properties for Biomedical Applications

Ángel Serrano-Aroca

Additional information is available at the end of the chapter

http://dx.doi.org/10.5772/intechopen.68996

Abstract

Acrylic-based polymers have many currently important biomedical applications such as contact lenses, corneal prosthesis, bone cements, tissue engineering, etc. due to their excellent biocompatibility and suitable performance in mechanical properties, among many other applications. Many of these biomaterials have been approved by the US Food and Drug Administration (FDA) for various applications. However, the potential uses of these polymeric materials in the biomedical industry could be increased exponentially if some of their acrylic properties (mechanical strength, electrical and/or thermal properties, water sorption and diffusion, biological interactions, antibacterial activity, porosity, etc.) are enhanced. Thus, acrylics have been fabricated as multicomponent polymeric systems in the form of interpenetrated polymer networks or combined with other advanced materials such as fibers, nanofibers, graphene and its derivatives and/or many other kinds of nanoparticles to form composite or nanocomposite materials, which are expected to exhibit superior properties. Besides, in regenerative medicine, acrylic scaffolds need to be designed with the required extent and morphology of pores by sophisticated techniques. Even though the great advances have been achieved so far, much research has to be carried out still in order to find new strategies to improve the above-mentioned properties.

Keywords: biomedical applications, hydrogels, acrylic properties, nanocomposites, latest improvements

1. Introduction

Acrylics are currently used in many important fields of the biomedical industry such as corneal prosthesis, intraocular lenses and contact lenses in ophthalmology [1], bone cements for orthopedic applications [2], tissue engineering [3], etc. due to their excellent properties such as biocompatibility and suitable mechanical performance, among others [4]. Many of these acrylic products for various applications have been approved by the US Food and Drug Administration (FDA) and are expected to produce massively. However, many of their potential uses required for many biomedical applications are sometimes hindered by their low mechanical strength, biological interactions, electrical and/or thermal properties, water sorption and diffusion, antibacterial activity, porosity, etc. when they are synthesized as scaffolds for tissue engineering applications. Thus, new advanced acrylic-based materials have been developed and are currently under intensive research to solve all these problems by means of multicomponent polymeric systems or by combination with other materials and/or nanomaterials to form composites or nanocomposites with or without interconnected porous morphology.

2. Mechanical properties

The improvement of the mechanical properties of acrylics is one of the *hot topics* in the field of bioengineering, and many research groups have been working on this topic for many years. Acrylics can be reinforced through many kinds of methods and techniques: polymers with microphase-separated morphologies such as block copolymers, in which hydrophobic and hydrophilic domains alternate [5], increasing crosslinking density [6], by means of binary systems composed of two or more mixed polymers as interpenetrating polymer networks (IPNs) [7], self-reinforced composite materials composed of fibers embedded in a matrix of the same acrylic polymer [8], by plasma grafting of a hydrophilic acrylic polymer onto a hydrophobic acrylic substrate [9, 10] and with the sol-gel reaction to produce nanosilica reinforcement [11]. However, more recent studies have shown new procedures to improve the mechanical properties of acrylics with the incorporation of graphene (GN) (2010 Nobel Prize in Physics) and other carbon nanomaterials such as carbon nanotubes (CNT) [12]. Chemically modified graphenes (CMGs) such as graphene oxide (GO) [13, 14] or reduced graphene oxide (rGO) [15], have also been shown to be very good nanofillers to reinforce acrylics and improve many other properties, especially acrylic hydrogels which, in the swollen state, show very low mechanical properties.

2.1. IPNs

Acrylic-based interpenetrating polymer networks (IPN) have gained greater attention during last decades, mainly due to their biomedical applications as reinforced polymer networks. The use of an IPN, which consists of two separate but interwoven polymer networks, is a chemical procedure that is often used in polymer science to control, enhance and/or combine functional properties. These are advanced multicomponent polymeric systems of crosslinked polymer networks without any covalent bonds between them, where at least one of them is synthesized and/or crosslinked within the immediate presence of the other. It is important to differentiate between the six basic multicomponent polymeric structures (**Figure 1**).

If a crosslinker is present in the polymeric system, fully IPN [17] result, while in the absence of crosslinking, a network having linear polymers embedded within the first crosslinked network is formed (semi- or pseudo-IPN) [18, 19]. Acrylic-based IPNs are prepared usually in the form

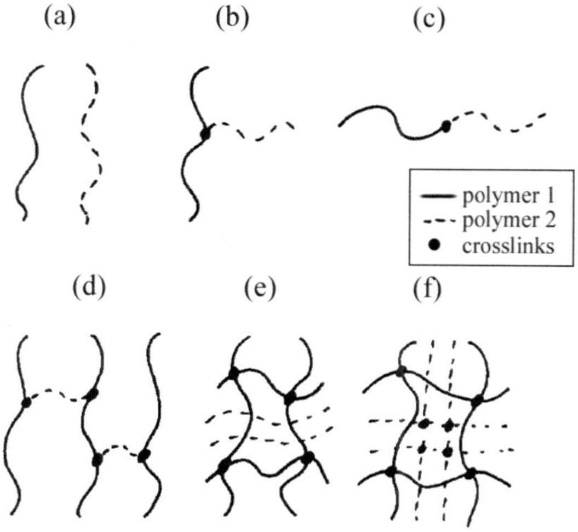

Figure 1. Schematic representation of (a) mechanical blends, (b) graft copolymers, (c) block copolymers, (d) AB-crosslinked copolymer, (e) semi-IPNs and (f) full-IPNs. Modified from Sperling and Mishra [16].

of simultaneous interpenetrating polymer networks (SINs), in which the precursors of both networks are mixed and the two networks are synthesized at the same time, or in the form of sequential IPNs, by swelling of a single-polymer network into a solution containing the mixture of monomer, initiator and activator, usually with a crosslinker. Thus, urethane acrylate resin networks were greatly reinforced by epoxy networks in SINs composed of both resins [20]. Full IPNs and semi-IPNs of the epoxy resin and poly(ethyl methacrylate) (PEMA) were also prepared by the sequential mode of synthesis, and these showed a gradual decrease of modulus and tensile strength properties with consequent increase in elongation at break and toughness for both types of IPNs with increases in PEMA content [21]. Combinations of different kinds of IPNs have been synthesized using simultaneous photopolymerization, which gave rise to simultaneous semi-interpenetrating polymer networks (semi-SINs) of epoxy resin-acrylate polyurethane semi-interpenetrating networks having very high compatibility [22].

Pseudo-SIPNs were prepared by melt blending of poly(methyl methacrylate) (PMMA), and double-C_{60}-end-capped poly(ethylene oxide) (FPEOF) exhibited a storage modulus of 16 times larger than that of PMMA, which are as good as those of PMMA/carbon nanotube nanocomposites [19].

Acrylic-based IPN hydrogels are also developed with the aim of enhancing the mechanical strength and swelling/deswelling response of these acrylic hydrogels [23]. For example, interpenetrating polymer network (IPN) hydrogels composed of chitosan and poly(acrylic acid) (PAA) are synthesized by employing the UV irradiation method and showed that even in the swollen state, the present chitosan/PAA IPNs possessed good mechanical properties [24].

'Smart' hydrogels are able to significantly change their volume/shape in response to small alterations of certain parameters of the environment. These responsive hydrogels have numerous applications, most of them focussing on biological and therapeutic demands [25, 26] and sensing applications [27].

Although IPNs based on hydrogels have been extensively reported, the combination of liquid crystalline (LC) property-based hydrogels has been rarely explored. In this case, the anisotropic and molecular order of liquid crystals can be combined with the responsive isotropic properties of hydrogels. Thus, advanced stimuli-responsive materials based on interpenetrating liquid crystal-hydrogel polymer networks have been recently fabricated consisting of a cholesteric liquid crystalline network that reflects color and an interwoven poly(acrylic acid) network that provides a humidity and pH response [28].

2.2. Composites

Several kinds of chemical modifications of acrylic hydrogels do not have a significant change of the overall mechanical strength because the main structural skeletons of these polymers or copolymers are still weak. On the contrary, the method of fiber reinforcement to produce composites is different because the added fabrics impart high strength to the networks which not only just embed inside the membranes but also form the main skeleton of the composites. A fiber-reinforced polymer is a composite material consisting of a polymer matrix imbedded with high-strength fibers such as glass, aramid and carbon [29]. In such kind of materials, the mechanical properties are presumed to be improved and the biocompatible characteristics of the acrylic polymer should remain the same. Thus, acrylic resin polymers have been reinforced with glass fibers for dental applications [30], and acrylic hydrogels such as poly(2-hydroxyethyl methacrylate), which is one of the most popular biomaterials, have been manufactured by adding various kinds of weaved and knitted fabrics and fibers, in order to improve overall qualities of the poly(2-hydroxethyl methacrylate) (PHEMA)-based artificial skin for advanced wound dressing usage [31]. However, in the recent decades, natural fibers as an alternative reinforcement in polymer composites have attracted the attention of many research groups due to their advantages over conventional glass and carbon fibers [32]. These natural fibers include flax, hemp, jute, sisal, kenaf, coir, kapok, banana, henequen and many others, which offer various advantages over man-made glass and carbon fibers such as low-cost, low-density, comparable specific tensile properties, non-abrasive to the equipment, non-irritation to the skin, reduced energy consumption, less health risk, renewability, recyclability and biodegradability [33]. Thus, ultra-long chitin natural fibers were incorporated into PMMA resin to prepare PMMA/chitin nanocomposites with improved properties [34]. This achievement is a significantly environmental move toward the sustainable utilization of marine-river crab shell wastes for biomedical applications.

2.3. Nanocomposites

Another alternative and very promising way of reinforcing acrylic polymers consists of the incorporation of nanomaterials such as silica, graphene and its derivatives, nanofibers or many other nanoparticles. Silica is a biocompatible material and has been reported to possess bioactive properties [35]. Silica can improve the mechanical properties of acrylics through nanosilica filling or the well-known sol-gel process, which offers a new approach to the synthesis of nanocomposite materials with domain sizes approaching the molecular level [36].

Thus, a biphasic matrix of a hybrid (inorganic-organic) nanocomposite materials of poly(2-hydroxyethyl acrylate) with a silica network obtained by an acid-catalyzed sol-gel process of tetraethoxysilane (TEOS) showed a very significant improvement of the mechanical properties of the pure hydrogel [37].

The combination of the reinforcement through interpenetrated polymer networks and nano-silica filling is another strategy that has also been used in the past. For example, simultaneous polyurethane/poly(ethyl methacrylate) interpenetrating polymer network with silica filler consisting of very fine powders with an approximate diameter of 5 nm also showed an important improvement of material strength [38].

Graphene (GN) is a two-dimensional (2D) monolayer of sp^2-bonded carbon atoms, which has attracted increasing attention [39] owing to its excellent electrical and thermal conductivities [40, 41] and great mechanical strength [42]. Besides, graphene promotes adherence of human osteoblasts and mesenchymal stromal cells [43], which render this nanomaterial also very promising material in the biomedical field. Thus, it has shown potential applications in nanocomposites such as magnetite-GNs/poly(arylene-ether-nitrile) nanocomposites because their mechanical properties were significantly enhanced by the incorporation of magnetite-GNs hybrids [44]. Besides, GN enhances the shape memory of poly(acrylamide-co-acrylic acid) and self-healing ability when the content of graphene is in the range of 10–30%, even though this copolymer itself has poor shape memory ability [45]. There are some reports that graphene oxide (GO) nanosheets can also enhance the mechanical strength of polymer substrates such as poly(acrylamide) (PAM) hydrogels [46]. GO is also a 2D nanomaterial prepared from natural graphite that can be easily exfoliated into monolayer sheets. GO has many hydrophilic oxygenated functional groups, including hydroxyl (-OH), epoxy (-C-O-C-), carbonyl (-C=O) and carboxyl (-COOH), which enable its dispersion in water solution [47]. The diversity of unique properties of GO, including great tensile modulus (1.0 TPa), ultimate strength (130 GPa) and electrical and thermal properties [48], renders graphene oxide an ideal carbon nanomaterial for variety of applications toward the development of new advanced materials. Thus, GO added into PAM hydrogels improved very much the mechanical performance of the original PAM hydrogels, which generally exhibit pronounced weakness and brittleness [46]. In the same way, the addition of GO nanosheets increased the Young's modulus and maximum stress of poly (acrylic acid)/gelatin composite hydrogels significantly as compared with control (0.0 wt.% GO). The highest Young's modulus was observed for hydrogel with GO (0.2 wt.%)/PAA (20 wt.%), whereas the highest maximum stress was detected for GO (0.2 wt.%)/PAA (40 wt.%) specimen. These results suggested that the application of GO nanosheets could be used to improve mechanical properties of hydrogel materials, which is very beneficial for tissue engineering applications [14]. The other derivatives of graphene such as chemically modified graphene (CMG) fillers have been used in nanocomposites of PMMA and were compared with the GO filling. These results showed an elastic modulus of GO/PMMA and RG-O/PMMA composites improved by 28% by just loading 1 wt.%. Fracture strength increased for GO/PMMA composites but decreased for RG-O/PMMA composites [49].

Single-wall carbon nanotubes (SWCNTs), multi-wall carbon nanotubes (MWCNTs) as well as carbon nanofibers (CNFs) are being used for reinforcing polymer matrices such as poly(methyl methacrylate) by melt blending. Thus, for example, using an amount of carbon nanofibers of 5 wt.%, the nanocomposites improved over 50% of their axial tensile modulus as compared to the control PMMA. The PMMA/CNFs nanocomposite fibers also show enhanced thermal stability, significantly reduced shrinkage and enhanced modulus retention with temperature, as well as improved compressive strength [50]. The electrical properties and electromechanical responses of acrylic materials such as acrylic elastomers and styrene copolymers can be improved toward electroactive applications such as artificial muscle and/or micro-electromechanical systems (MEMS) devices [51].

The other novel nanocomposite hydrogels such as those prepared with polyacrylamide (PAM) as a matrix material reinforced with natural chitosan nanofibers via *in situ* free-radical polymerization showed that these nanofibers acted as a multifunctional crosslinker and a reinforcing agent in the hydrogel system producing a compression strength and a storage modulus significantly higher than those of pure PAM [52].

Reinforcement can be also performed with plant fiber-based nanofibers by a successful fibrillation of wood pulp fibers into nanofiber bundles, which are thin enough to work, as well as bacterial cellulose in maintaining the transparency of resin [53].

The other nanoparticles such as clay have been employed to reinforce acrylic polymers. These polymer-clay nanocomposites such as PMMA/clay constitute a class of materials in which the polymer matrix is reinforced by uniformly dispersed inorganic particles (usually 10 wt.% or less) having at least one dimension in the nanometre scale and exhibiting enhanced mechanical and thermal properties when compared to pure polymer or conventional composites [54].

3. Electrical properties

The electrical properties are very important in some biomedical fields because various types of electrical stimulation can regulate cell physiological activities such as division [55], migration [56], differentiation and cell death [57]. The electrical stimulation also has been employed in promoting healing for spinal cord repair and cancer therapy due to its non-invasiveness of these polymers [58–60]. For these reasons, much emphasis was given in developing these new acrylic-based materials in biomedical application where the conductivity of the biomaterials is essential. Recently, nanocarbon materials such as graphene had been considered to be very effective electrode material with very high conductivity. The graphene has high transmittance and excellent conductivity [40] as was mentioned earlier. But, its production is still very expensive, and more studies are expected to carry out with its derivative, graphene oxide. However, in order to develop electrically conductive acrylic-based resins, GO, which has a very low conductivity due to their oxygen-containing functional groups, must be modified to obtain reduced graphene oxide (rGO). Thus, for example, following a single-step procedure starting from a homogeneous water dispersion of GO, it is possible to undergo reduction induced by the UV radiation during the photopolymerization of an acrylic resin [61]. The transparent conductive films were also produced by grafting poly(acryl amide)/poly(acrylic acid) on the

GO surface followed by a reduction to rGO nanosheets by a two-step chemical reduction with increased conductivity [62]. The inorganic-organic double network (DN) flexible and conductive hydrogel of rGO and poly(acrylic acid) has also been prepared by a two-step synthesis with a reduction-induced *in situ* self-assembly [63]. Even more recently, a nacre-inspired acrylic-based nanocomposite of rGO and PAA has been prepared *via* a vacuum-assisted filtration self-assembly process (**Figure 2**). The abundant hydrogen bonding between GO and PAA results in both high strength and toughness of the bioinspired nanocomposites, which are 2 and 3.3 times higher than that of pure reduced GO film, respectively. Moreover, this nanocomposite also displays high electrical conductivity of 108.9 $S{\cdot}cm^{-1}$, which renders it very promising material in many biomedical applications such as flexible electrodes, artificial muscles, etc.

Carbon nanotubes (CNTs), discovered by Iijima [65], have also been attracting intensive attention because of their excellent electrical properties with a superb conductivity, remarkable mechanical strength and modulus with many potential technological applications [66]. CNTs offer the possibility of developing ultrasensitive electrochemical biosensors due to unique electrical properties. Thus, nanofibrous membranes filled with multi-walled carbon nanotubes (MWCNT) were electrospun from the mixture of poly(acrylonitrile-co-acrylic acid) (PANCAA) and MWCNT to develop a glucose biosensor for diabetics [67].

The other nanocomposites of poly(methyl methacrylate) containing various multi-walled carbon nanotube (MWCNT) contents have been prepared using melt mixing to achieve high conductivity levels in the nanocomposites [68].

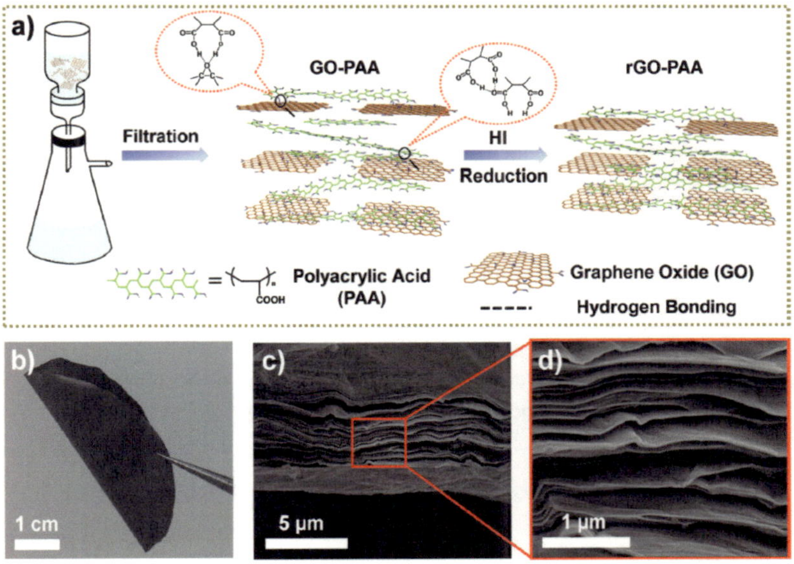

Figure 2. Fabrication process of rGO–PAA nanocomposites: (a) The GO nanosheets/PAA homogeneous solution was filtered by vacuum-assisted filtration into GO–PAA nanocomposites. Then after HI reduction, the rGO–PAA nanocomposites were obtained. (b) A digital photograph of rGO–PAA nanocomposites. (c) and (d) The cross-section surface morphology with different magnifications of rGO–PAA nanocomposites. Reprinted with permission from Ref. [64].

4. Thermal properties

Even though biomaterials do not need to endure temperatures higher than that of the human body, the improvement of thermal properties can increase its long-term operation. Thus, for example, the incorporation of polyurethane into polyacrylamide network in the form of an interpenetrating polymer networks enhanced the thermal properties of these semi-IPNs due to higher crosslink density imparted by the hard segment content [18]. Though silica can improve the mechanical properties of acrylic polymers, the differential scanning calorimetry results of PHEMA/SiO_2 hybrids are complicated, showing two glass transition temperatures, and it was shown that the SiO_2 content is an important factor in influencing the shift of the Tg transition [69]. However, polymer nanocomposites with functionalized graphene sheets (FGNS) showed an unprecedented shift in glass transition temperature of up to 40 and 30°C in poly(acrylonitrile) with 1 wt.% of FGNS and in poly(methyl methacrylate) with only 0.05 wt.%, respectively [70]. Besides, the thermal stability of magnetite-graphene/poly(arylene-ether-nitrile) nanocomposites were significantly enhanced by the incorporation of magnetite-graphene hybrids [44]. The nanocomposites of PMMA with chemically modified graphene (CMG) and GO fillers made by *in situ* polymerization showed large shifts in the glass transition temperature with loadings as low as 0.05 wt.% [49].

Another strategy to improve the thermal properties of acrylic polymers is by nanoparticle filling. Thus, the thermal performance of well-known acrylic polymers such as PMMA can be significantly improved by filling of nanometric particles (5, 10 15 and 20%) of titanium oxide (TiO_2) and ferric oxide (Fe_2O_3) by the solvent casting method [71, 72].

Thermal degradation can also be improved in acrylic-based materials by nanoparticle filling. For example, the experimental results obtained by thermogravimetric analysis (TGA) of PMMA with TiO_2 and Fe_2O_3 showed that these nanoparticles can improve the thermal stability of PMMA by about 50°C by loading 5 wt.% of fillers [72]. The TGA also showed that the presence of small amounts of Pd nanoparticles (0.0005–0.005 vol%) in PMMA/Pd nanocomposites significantly improved the thermal stability of PMMA, as shown by a degradation initiation retarded by 75°C and a gain of 32°C at the maximum decomposition rate [73].

Acrylic hydrogels are hydrophilic polymers and are able to absorb large amounts of water in their biomedical applications due to contact with cells or tissue in the human body. Therefore, the thermal analysis of water and its influence on the swollen hydrogel properties becomes essential. Thus, many studies have been done in this way with acrylic hydrogels such as PHEMA [74], bulk and plasma-polymerized poly(2-hydroxyethyl acrylate) (PHEA) [75], Poly(ethyl acrylate) [76], etc.

5. Water sorption and diffusion

Water sorption and diffusion are also very important in biomedicine because these properties play a very important role in cell survival, especially in tissue engineering applications [3].

Thus, acrylic hydrogels such as poly(2-hydroxyethyl methacrylate) or poly(2-hydroxyethyl acrylate), are very important hydrophilic materials as these polymers were able to absorb and swell retaining large amounts of water within their structure[77–80]. The excellent water sorption property has made these types of materials very promising in a wide range of biomedical applications such as controlled drug delivery, tissue engineering, wound healing, etc. [4, 81]. The ability of hydrogels to absorb water arises from hydrophilic functional groups attached to the polymeric backbone, while their resistance to dissolution arises from crosslinks between network chains [82]. However, these single-network hydrogels have weak mechanical properties and slow response at swelling. Therefore, they are in need of reinforcement, as already mentioned, which can also modify their water sorption properties. For example, the combination of hydrophilic and hydrophobic functional groups of acrylic polymers as multicomponent polymeric systems is shown in **Figure 1**.

The reinforcement of acrylics through GO loading can modify the water sorption behavior of the polymers. Thus, the swelling rates of graphene oxide/poly(acrylic acid-co-acrylamide) nanocomposite hydrogels increased with increase in the GO loadings to 0.30 wt.% and then decreased with further increasing GO loadings. It is worth noting that the hydrogel with only 0.10 wt.% GO exhibited significant improvement of swelling capacity in neutral medium and could also retain relatively higher swelling rates to a certain degree in acidic and basic solutions. Therefore, these GO-based superabsorbent acrylic hydrogels have very potential applications in many fields such as biomedical engineering and hygienic products [47].

The mechanism of water diffusion [83] can also be altered by the reinforcement of acrylics through any of the methods shown in Section 1. Thus, poly(acrylic acid)-GO nanocomposite hydrogels shows non-Fickian anomalous diffusion and the deswelling ratio decreases with increasing GO content [48].

A new method (ultrasound synthesis) has been developed to prepare superabsorbent polymers of sodium lignosulfonate-grafted poly(acrylic acid-co-acryl amide). This superabsorbent acrylic-based polymer exhibited also a non-Fickian water diffusion transport and a maximum water absorbency of 1350 $g \cdot g^{-1}$ [84].

There are many acrylic hydrogels, which exhibit a non-Fickian diffusion behavior such as poly(2-hydroxyethyl acrylate) [79, 80]. Even though water sorption is not classically Fickian, it has been observed in a variety of polymers such as PHEMA, that an important water-swellable biomedical polymer is controlled by Fickian diffusion [85]. Thus, copolymeric hydrogels based on 2-hydroxyethyl methacrylate (HEMA) and epoxy methacrylate (EMA) synthesized by bulk polymerizations showed that the swelling process of these polymers also follows Fickian behavior and the equilibrium water content (EWC) decreased with increase in EMA content due to its hydrophobicity [86].

It is remarkable that the pH has a big influence in the swelling properties and diffusion mechanism of acrylic-based materials. Thus, the swelling properties of semi-interpenetrating polymer networks of acrylamide-based polyurethanes decreased in acidic pH, while a reverse trend was seen in basic pH. However, these semi-IPNs were found to be hydrolytically stable in phosphate buffer solution, which makes them to be a potential material for biomedical applications [18].

Polyacrylic acid is a pH-sensitive and biocompatible polymer that is being used in many biomedical fields [26]. It has attracted considerable interest because of its therapeutic use, due to its ability to swell reversibly with changes in pH. Thus, GO functionalized with PAA (GO-PAA) by *in situ* atom transfer radical polymerization (ATRP) showed potential use as an intracellular protein carrier using bovine serum albumin (BSA) as a model protein [87]. This application is very important because proteins participate in all vital body processes and these perform an essential function inside cells as enzymes, transduction signals and gene regulation. Another pH-sensitive terpolymer hydrogel, poly(acryl amide-co-2-acrylamido-2-methyl-1-propanesulfonic acid-co-acrylamido glycolic acid), with applications in drug release showed a quasi-Fickian diffusion mechanism with partly chain relaxation controlled diffusion. These hydrogels demonstrated a sharp change in its water absorbency and molecular weight between crosslinks of the network with a change in pH of the swelling media [88].

The effect of temperature on swelling properties of acrylic hydrogels is also very important [86], and they can be modified to exhibit fast temperature sensitivity, and improved oscillating swelling-deswelling properties as, for example, in thermosensitive poly(N-isopropyl acrylamide-co-acrylic acid) hydrogels [89].

6. Antibacterial activity

In biomedicine, bacterial infections can lead to implant failure, which may cause major economic losses and suffering among patients despite the use of preoperative antibiotic prophylaxis and the aseptic processing of materials. Therefore, novel antibacterial materials are urgently needed for medical uses [90]. However, acrylics itself do not have antibacterial activity intrinsically, and therefore some fillers and antibacterial agents need to be incorporated by physical blending in order to produce an acrylic-based antibacterial material [91]. Thus, graphene has emerged as a novel green broad-spectrum antibacterial material, with little bacterial resistance and tolerable cytotoxic effect on mammalian cells. It exerts its antibacterial action via physical damages through direct contact of its sharp edges with bacterial membranes and destructive extraction of lipid molecules. The graphene-based nanocomposites have a wide range of biomedical applications such as wound dressing due to its superior antibacterial properties and good biocompatibility [92].

In the field of dental materials, since methyl methacrylate was firstly used in tooth restoration in 1937, methacrylate monomers having good biocompatibility and adhesive property have been extensively used as dental materials [91]. The most commonly used methacrylate monomers in commercial dental resin-based materials are methyl methacrylate (MMA), 2,2-bis[4-(2-hydroxy-3-methacryloyloxypropyl)-phenyl]propane (Bis-GMA), 1,6-bis-[2-methacryloyloxyethoxycarbo nylamino]-2,4,4-trimethylhexane (UDMA) and tri-ethylene glycol dimethacrylate (TEGDMA) [93]. However, the dental materials produced with these monomers are not of antibacterial nature, which is very important in this biomedical field. Thus, another strategy to design acrylic hydrogels with desired antibacterial performance consists of adding silver nanoparticles (Ag NPs). This modification produced a strong antibacterial activity against *Escherichia coli* and also improved the mechanical properties of acrylic resins for dental applications. Such antibacterial effects were mainly attributed to the release of silver ions upon immersion of the dental composite in water, which appeared to be fairly nontoxic to humans [94]. Poly(methyl methacrylate) (PMMA) nanofibers containing silver nanoparticles were synthesized by radical-mediated dispersion polymerization and also showed enhanced antimicrobial efficacy compared to that of silver sulfadiazine and silver nitrate at the same silver concentration [95].

Infections are also frequent and highly undesired occurrences after orthopedic procedures. Besides, the growing concern caused by the rise in antibiotic resistance progressively decreased the efficacy of such drugs. Thus, in this area, the integration of silver nanoparticles in the polymeric mineralized acrylic-based nanocomposites also provides antibacterial activity against bacteria [96].

The combination of both previous strategies (graphene and Ag NPs) to design antibacterial hydrogels with good water-maintaining ability is of particular significance to promote the development of wound dressing. Thus, a series of hydrogels were synthesized by crosslinking of Ag/graphene composites with acrylic acid and N,N′-methylene bisacrylamide at different mass ratios. In this study, prepared hydrogel with an optimal Ag to graphene mass ratio of 5:1 exhibited much stronger antibacterial abilities than other hydrogels and showed excellent biocompatibility, high swelling ratio and good extensibility at the same time. Besides, *in vivo* experiments indicated that this nanocomposite hydrogel could significantly accelerate the healing rate of artificial wounds in rats, and it helped to successfully reconstruct intact and thickened epidermis during 15 days of healing of impaired wounds [97]. In the same way, acrylic acid (AA) grafted onto poly(ethylene terephthalate) (PET) film through gamma ray-induced graft copolymerization with silver nanoparticles on the surface showed strong and stable antibacterial activity [98].

7. Porosity in scaffolds for tissue engineering

Tissue engineering holds great promise for regeneration and repair of diseased tissues, making the development of new porous supports as scaffolds for tissue regeneration a topic of great interest in biomedical research. Hydrogels have emerged as leading candidates for engineered tissue scaffolds due to of their biocompatibility and similarities to native extracellular matrix. However, precise control of hydrogel properties such as high porosity, remains a challenge. Traditional techniques for creating bulk porosity in polymers have demonstrated success in hydrogels for tissue engineering. However, some problems related to direct cell encapsulation often occur. Emerging technologies have demonstrated the ability to control porosity and morphology in hydrogels, creating engineered tissues with structure and function similar to native tissues [99].

The applications of porous materials are nowadays widespread. The interconnection and geometry of pores, which depend on the tissue to regenerate, physicochemical properties and mechanical resistance of the material, play in these biomedical applications a major role. Thus, there are several methods to produce *scaffolds*, which include gas foaming [100], sintering fiber meshes [101], solvent casting [102], polymerization in solution [80, 103], porogen technique [104, 105], freeze-drying techniques [106, 107], electrospinning [108], 3D printing [109], 3D bioplotting of scaffold with cells [110], etc. For example, acrylic scaffolds with interconnected spherical pores and controlled hydrophilicity with interconnected porous structure were synthesized using a template of sintered PMMA microspheres of controlled size. In these scaffolds, the geometric characteristics (pore size, connectivity and porosity) and the physicochemical properties of the resulting material can be controlled in an independent way. Copolymerization of hydrophobic ethyl acrylate and hydrophilic hydroxyethyl methacrylate comonomers in the free space of the template and subsequent solution of the PMMA microspheres gave rise to the scaffold with the designed pore architecture (see **Figure 3**) [105].

Another example is the novel preparation of gelatin-PHEMA porous scaffolds by freeze-drying technique, in which morphology was assessed by SEM and μ-CT (**Figure 4**). Four types of novel hydrogels using different methacrylamide-modified gelatin/2-hydroxyethyl methacrylate ratios between 1/0 and 1/2 (w/w) (samples from C0 to C3) were prepared in this study, and the results indicated that the HEMA content in the initial polymerization mixtures modulates the architecture of the porous scaffolds from straightforward, top-to-bottom oriented channels for hydrogels possessing the lowest HEMA content to a complex and dense internal porosity of the channels in the case of higher HEMA loaded materials. It is important to notice that the covalently bound gelatin sequences significantly improve the biocompatibility of PHEMA-based hydrogels, which is very convenient for tissue engineering purposes.

Superporous acrylic scaffolds can also be prepared by the salt-leaching technique using NaCl or $(NH_4)_2SO_4$ as a porogen [111] or with many other porogenic agents such as ammonium oxalate crystals [112].

By submitting carbon dioxide (CO_2) to supercritical conditions (P = 160–260 bar, T = 60°C) after certain time and then rapidly depressurize, it is also possible to fabricate porous structures that are related to the supercritical parameters and to the polymer blend composition [113]. The use of carbon dioxide (CO_2) to create such scaffolds has received some attention in the past. But, many researchers believe that although CO_2 processing of polymers can lead to porous scaffolds, there is limited interconnectivity between the pores. However, highly porous (greater than 85%) and well-interconnected scaffolds were obtained in a blend of poly(ethyl methacrylate) and tetrahydrofurfuryl methacrylate (PEMA/THFMA) showing promise for potential applications in cartilage repair [114].

Probably the most sophisticated techniques to produce scaffolds are electrospinning, 3D printing and bioprinting. Electrospinning is composed of a high-voltage DC power supply, an infusion pumps and a syringe with a needle tip usually with a diameter of 0.5 mm, and for example, biodegradable nanofibrous poly(L-lactic acid) (PLLA) scaffolds have been prepared by this process for use in tissue regeneration [115]. 3D printing promises to produce complex biomedical devices according to computer design using patient-specific anatomical data. This 3D printing technique has slowly evolved to create one-of-a-kind devices, implants, scaffolds for tissue engineering and drug delivery systems among other important applications. However, several technological limitations, related to the kind of commercially printable materials available and other technical printing aspects, must still be overcome. The common 3D printing technologies such as three-dimensional printing, fused deposition modeling, selective laser sintering, stereolithography and 3D Plotting/Direct-Write/Bioprinting, are still

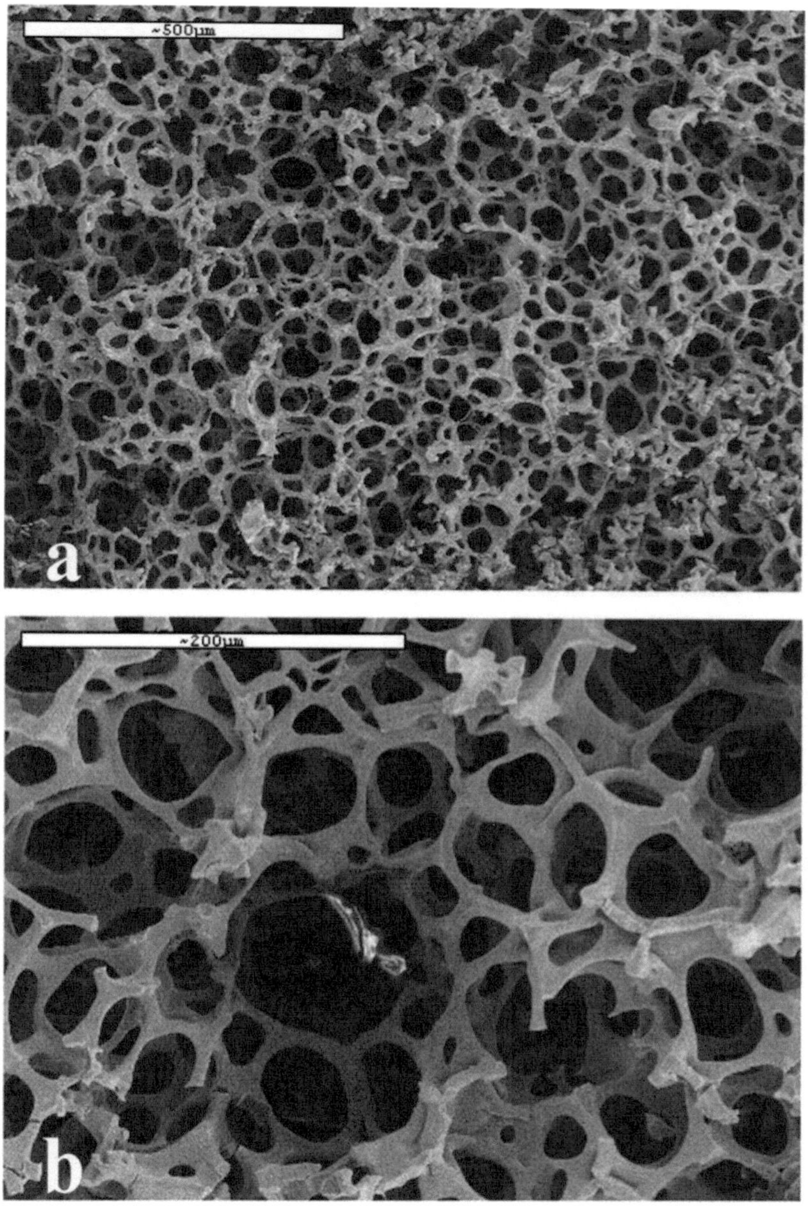

Figure 3. SEM micrographs of EA/HEMA copolymer scaffolds (30% HEMA) at different magnifications. Reprinted with permission from Ref. [105].

under deep research for the progress of each technology in tissue engineering. Bioprinting is the more advanced 3D printing technology because it consists of printing cells combined with custom 3D scaffolds for personalized regenerative medicine [109].

Mechanical resistance depends both on the material properties and on the interconnected pore structure of the scaffold. This problem is more serious in the case of porous acrylic hydrogels, which exhibit very poor mechanical properties in their swollen state [80]. Therefore, it is

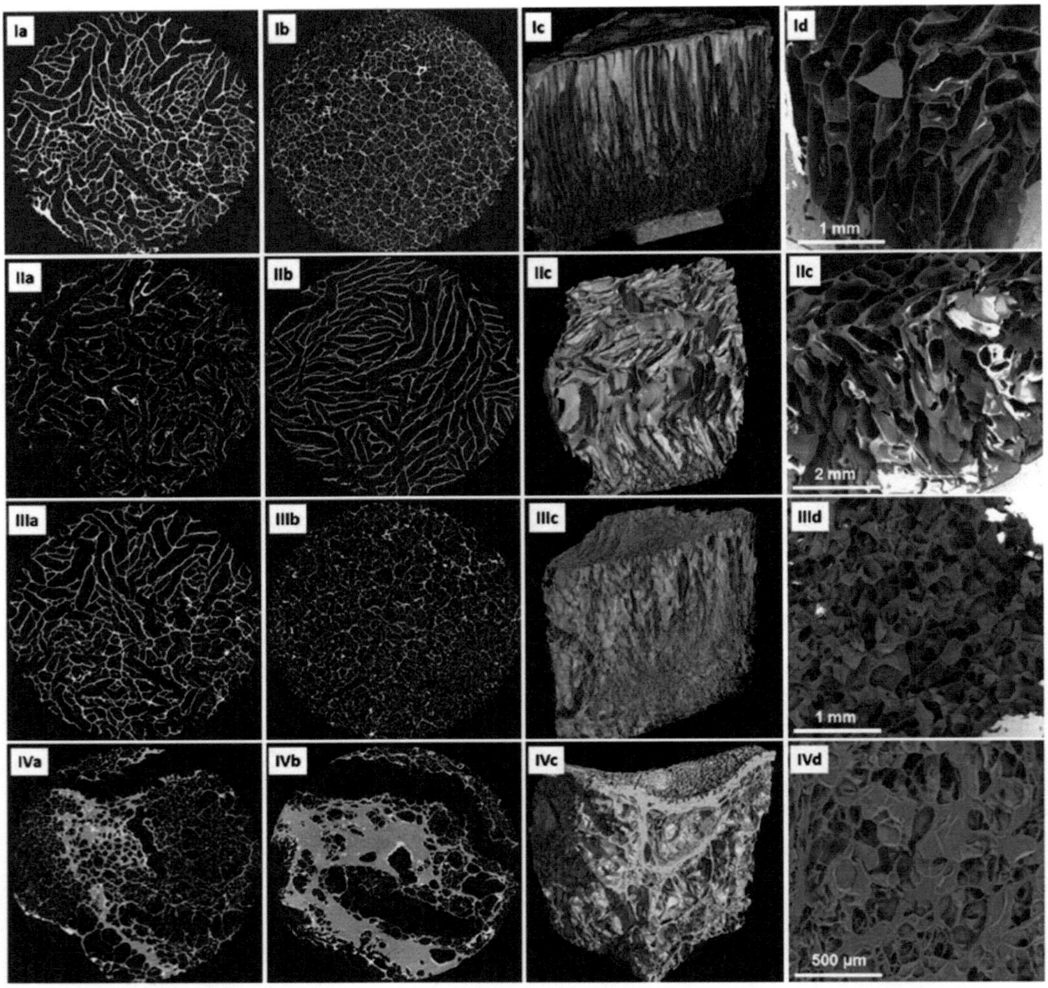

Figure 4. Morphology of the gelatin-PHEMA porous scaffolds as obtained through μ-CT (panels (a)–(c): (a), top view; (b), bottom view; (c), side view) and SEM (panel (d)) analyses: panel (I), C0; panel (II), C1; panel (III), C2; panel (IV), C3. Reprinted with permission from Ref. [107].

usually necessary to enhance the mechanical properties of these porous structures by means of the methods, shown in Chapter 2, with nanomaterials or other techniques. For example, the use of a hybrid hydrogel nanocomposite of silica/poly(2-hydroxyethyl acrylate) as scaffold material matrix greatly improves the mechanical properties, and the silica phase of the scaffold was effectively interconnected and continuous, able to withstand pyrolysis without losing the pore architecture of the scaffold [37].

Other modifications of acrylics such as those of PHEMA with cholesterol methacrylate (CHLMA) and laminin have been developed in the presence of ammonium oxalate crystals to introduce interconnected superpores in the matrix in order to design superporous scaffolds that promote cell-surface interaction [116]. PHEMA has also been modified with laminin-derived Ac-CGGASIKVAVS-OH peptide sequences to construct scaffolds that promote cell adhesion and neural differentiation. With the same goal, nanofiber scaffolds prepared by electrospinning were treated with oxygen plasma and then simultaneously *in situ* grafted with hydrophilic acrylic acid to obtain PLLA-g-PAA with a modified surface, which significantly improved cell adhesion and proliferation [115].

Recent studies of 3D microenvironment comprising fibronectin-coated PMMA/PC-based multicomponent polymeric systems promoted differentiation of primary human osteoblasts, which hereby renders a promising tool for tissue-specific *in vitro* preconditioning of osteoblasts designated for clinically oriented bone augmentation or regeneration. Furthermore, morphogenesis and fluorescence dye-based live/dead staining revealed homogenous cell coverage of the microcavities, whereas cells showed high viability up to 14 days, and Azur II staining proved formation of uniform sized multi-layered aggregates, exhibiting progressive intracellular deposition of extracellular bone matrix constituents comprising fibronectin, osteocalcin and osteonectin from day 7 onwards [117].

Polysaccharide hydrogels have become increasingly studied as matrices in soft-tissue engineering due to their known cytocompatibility. Thus, for example, crosslinkable dextran methacrylates and hyaluronan methacrylate hydrogels, which are candidates as matrices for soft-tissue reconstruction, were synthesized, showing that the *in vitro* degradation behavior of these kinds of hydrogels could be controlled by the polysaccharide structure and the crosslinking density. Furthermore, under *in vitro* conditions, these materials had no cytotoxic effects against fibroblasts, and the use of composite gels improved the adherence of cells [118].

Even though the great advances have been achieved in scaffold design of acrylic-based materials, much research has to be conducted still in order to find new ways and methods capable of providing suitable advanced porous materials for tissue engineering applications.

Acknowledgements

This work was supported by the 2015-162-001UCV grant from the Universidad Católica de Valencia *San Vicente Mártir*.

Author details

Ángel Serrano-Aroca

Address all correspondence to: angel.serrano@ucv.es

Bioengineering and Cellular Therapy Group, Department of Applied and Technological Sciences, Faculty of Veterinary and Experimental Sciences, San Vicente Mártir Catholic University of Valencia, Valencia, Spain

References

[1] Chirila T, Harkin D. Biomaterials and Regenerative Medicine in Ophthalmology. 2nd ed. UK, Elsevier; 2016

[2] Shalaby S, Nagatomi S, Peniston S. Polymeric Biomaterials for Articulating Joint Repair and Total Joint Replacement. Polymers for Dental and Orthopedic Applications; Boca Raton, CRC Press; 2007

[3] Van Blitterswijk C, De Boer J. Tissue Engineering. United States of America, Academic Press; 2014

[4] Ratner BD, Hoffman AS, Schoen FJ, Lemons JE. Biomaterials Science: An Introduction to Materials in Medicine. Canada, Academic Press; 2012

[5] Stoy V, Climent C. In Hydrogels: Speciality Plastics for Biomedical and Pharmaceutical Applications. Basel: Technomic Publishers; 1996

[6] Serrano-Aroca Á, Monleón-Pradas M, Gómez-Ribelles JL. Effect of crosslinking on porous poly(methyl methacrylate) produced by phase separation. Colloid and Polymer Science. 2008;**286**(2):209-216

[7] Ramaraj B, Radhakrishnan G. Modification of the dynamic swelling behaviour of poly(2-hydroxyethyl methacrylate) hydrogels in water through interpenetrating polymer networks (IPNs). Polymer (Guildf). 1994;**35**(10):2167-2173

[8] Gilbert JL, Ney DS, Lautenschlager EP. Self-reinforced composite poly (methy1 methacrylate): Static and fatigue properties. Journal of Applied Polymer Science 1995;**16**(14): 1043-1055

[9] Serrano-Aroca Á, Gómez-Ribelles JL, Monleón-Pradas M, Vidaurre-Garayo A, Suay-Antón J. Characterisation of macroporous poly(methyl methacrylate) coated with plasma-polymerised poly(2-hydroxyethyl acrylate). European Polymer Journal. 2007;**43**(10):4552-4564

[10] Serrano-Aroca Á, Monleón-Pradas M, Gómez-Ribelles JL. Plasma-induced polymerisation of hydrophilic coatings onto macroporous hydrophobic scaffolds. Polymer (Guildf). 2007;**48**(7):2071-2078

[11] Amerio E, Fabbri P, Malucelli G, Messori M, Sangermano M, Taurino R. Scratch resistance of nano-silica reinforced acrylic coatings. Progress in Organic Coatings, 2008;**62**:129-133

[12] Kugler S, Kowalczyk K, Spychaj T. Progress in organic coatings hybrid carbon nanotubes / graphene modified acrylic coats. Progress in Organic Coatings. 2015;**85**:1-7

[13] Cha C, Shin SR, Gao X, Annabi N, Dokmeci MR, Tang XS, et al. Controlling mechanical properties of cell-laden hydrogels by covalent incorporation of graphene oxide. Small. 2014;**10**(3):514-523

[14] Faghihi S, Gheysour M, Karimi A, Salarian R. Fabrication and mechanical characterization of graphene oxide-reinforced poly (acrylic acid)/gelatin composite hydrogels. Journal of Applied Physics. 2014;**115**(8)

[15] Ha H, Shanmuganathan K, Ellison CJ. Mechanically stable thermally crosslinked poly(acrylic acid)/reduced graphene oxide aerogels. ACS Applied Materials & Interfaces. 2015;**7**(11):6220-6229

[16] Sperling LH, Mishra V. The current status of interpenetrating polymer networks. Polymers for Advanced Technologies. 1996;**7**(4):197-208

[17] Maity J, Ray SK. Enhanced adsorption of methyl violet and congo red by using semi and full IPN of polymethacrylic acid and chitosan. Carbohydrate Polymers. 2014;**104**(1):8-16

[18] Merlin DL, Sivasankar B. Synthesis and characterization of semi-interpenetrating polymer networks using biocompatible polyurethane and acrylamide monomer. European Polymer Journal. 2009;**45**(1):165-170

[19] Wang M, Pramoda KP, Goh SH. Mechanical behavior of Pseudo-Semi-Interpenetrating polymer networks based on Double-C 60 -End-Capped poly (ethylene oxide) and Poly (methyl methacrylate). Chemistry of Materials. 2004;**16**:3452-3456

[20] Hua F, Hu C. Interpenetrating polymer networks of epoxy resin and urethane acrylate resin 2. Morphology and mechanical property. European Polymer Journal. 2000;**36**(1):27-33

[21] Chakrabarty D, Das B, Roy S. Epoxy resin-poly (ethyl methacrylate) interpenetrating polymer networks: Morphology, mechanical. 1997;**105**:1-9

[22] Vabrik R, Czajlik I, Tury G, Rusznak I, Ille A, Vig A. A study of epoxy resin-acrylated polyurethane semi-interpenetrating polymer networks. Journal of Applied Polymer Science. 1998;**68**(1):111-119

[23] Dragan ES. Design and applications of interpenetrating polymer network hydrogels. A review. Chemical Engineering Journal. 2014;**243**:572-590

[24] Lee J, Kim S, Kim S, Lee Y, Lee K. Synthesis and characteristics of interpenetrating polymer network hydrogel composed of chitosan and poly (acrylic acid). Polymer (Guildf). 1998;**73**:113-120

[25] Bajpai AK, Shukla SK, Bhanu S, Kankane S. Responsive polymers in controlled drug delivery. Progress in Polymer Science. 2008;**33**(11):1088-1118

[26] Zhao Y, Kang J, Tan T. Salt-, pH- and temperature-responsive semi-interpenetrating polymer network hydrogel based on poly(aspartic acid) and poly(acrylic acid). Polymer (Guildf). 2006;**47**(22):7702-7710

[27] Richter A, Paschew G, Klatt S, Lienig J, Arndt K-F, Adler H-JP. Review on Hydrogel-based pH sensors and microsensors. Sensors. 2008;**8**(1):561-581

[28] Stumpel JE, Gil ER, Spoelstra AB, Bastiaansen CWM, Broer DJ, Schenning APHJ. Stimuli-responsive materials based on interpenetrating polymer liquid crystal hydrogels. Advanced Functional Materials. 2015;**25**(22):3314-3320

[29] Groover MP. Fundamentals of Modern Manufacturing_ Materials, Processes, and Systems. 4th ed. J. Wiley & Sons; United States of America, 2010

[30] Stipho HD. Repair of acrylic resin denture base reinforced with glass fiber. Journal of Prosthetic Dentistry. 1998;**80**(5):546-550

[31] Young CD, Wu JR, Tsou TL. High-strength, ultra-thin and fiber-reinforced pHEMA artificial skin. Biomaterials. 1998;**19**(19):1745-1752

[32] Saheb DN, Jog JP. Natural fiber polymer composites: A review. Advances in Polymer Technology. 1999;**18**(4):351-363

[33] Ku H, Wang H, Pattarachaiyakoop N, Trada M. A review on the tensile properties of natural fiber reinforced polymer composites. Composites Part B: Engineering. 2011;**42**(4):856-873

[34] Chen C, Li D, Hu Q, Wang R. Properties of polymethyl methacrylate-based nanocomposites: Reinforced with ultra-long chitin nanofiber extracted from crab shells. Materials & Design. 2014;**56**:1049-1056

[35] Kokubo T. Design of bioactive bone substitutes based on biomineralization process. Materials Science and Engineering: C. 2005;**25**(2):97-104

[36] Hench LL, West JK. The sol-gel process. Chemical Reviews. 1990;**90**(1):33-72

[37] Rodríguez-Hernández JC, Serrano-Aroca Á, Gómez-Ribelles JL, Monleón-Pradas M. Three-dimensional nanocomposite scaffolds with ordered cylindrical orthogonal pores. Journal of Biomedical Materials Research Part B: Applied Biomaterials. 2008;**84**(2):541-549

[38] Trakulsujaritchok T, Hourston DJ. Damping characteristics and mechanical properties of silica filled PUR/PEMA simultaneous interpenetrating polymer networks. European Polymer Journal. 2006;**42**(11):2968-2976

[39] Novoselov KS, Geim AK, Morozov SV, Jiang D, Zhang Y, Dubonos SV, et al. Electric field effect in atomically thin carbon films. Science. 2004;**306**(5696):666-669

[40] Geim AK, Novoselov KS. The rise of graphene. Nature Materials. 2007;**6**:183-191

[41] Balandin AA, Ghosh S, Bao W, Calizo I, Teweldebrhan D, Miao F, et al. Superior thermal conductivity of single-layer graphene. Nano Letters. 2008;**8**(3):902-907

[42] Lee C, Wei X, Kysar JW, Hone J. Measurement of the elastic properties and intrinsic strength of monolayer graphene. Science. 2008;**321**(5887):385-388

[43] Kalbacova M, Broz A, Kong J, Kalbac M. Graphene substrates promote adherence of human osteoblasts and mesenchymal stromal cells. Carbon N Y. 2010;**48**(15):43234329

[44] Zhan Y, Meng F, Yang X, Liu X. Magnetite-graphene nanosheets (GNs)/poly(arylene ether nitrile) (PEN): Fabrication and characterization of a multifunctional nanocomposite film. Colloids and Surfaces A: Physicochemical and Engineering Aspects. 2011; **390**(1-3):112-119

[45] Dong J, Ding J, Weng J, Dai L. Graphene enhances the shape memory of poly (acrylamide-co-acrylic acid) grafted on graphene. Macromolecular Rapid Communications. 2013;**34**(8):659-664

[46] Shen J, Yan B, Li T, Long Y, Li N, Ye M. Study on graphene-oxide-based polyacrylamide composite hydrogels. Composites Part A: Applied Science and Manufacturing. 2012;**43**(9):1476-1481

[47] Huang Y, Zeng M, Ren J, Wang J, Fan L, Xu Q. Preparation and swelling properties of graphene oxide/poly(acrylic acid-co-acrylamide) super-absorbent hydrogel nanocomposites. Colloids and Surfaces A: Physicochemical and Engineering Aspects. 2012;**401**:97-106

[48] Shen J, Yan B, Li T, Long Y, Li N, Ye M. Mechanical, thermal and swelling properties of poly(acrylic acid)-graphene oxide composite hydrogels. Soft Matter. 2012;**8**(6):1831-1836

[49] Potts JR, Lee SH, Alam TM, An J, Stoller MD, Piner RD, et al. Thermomechanical properties of chemically modified graphene/poly(methyl methacrylate) composites made by *in situ* polymerization. Carbon N Y. 2011;**49**(8):2615-2623

[50] Zeng J, Saltysiak B, Johnson WS, Schiraldi DA, Kumar S. Processing and properties of poly(methyl methacrylate)/carbon nano fiber composites. Composites Part B: Engineering. 2004;**35**(2):173-178

[51] Kunanuruksapong R, Sirivat A. Electrical properties and electromechanical responses of acrylic elastomers and styrene copolymers: Effect of temperature. Applied Physics A: Materials Science & Processing. 2008;**92**(2):313-320

[52] Zhou C, Wu Q. A novel polyacrylamide nanocomposite hydrogel reinforced with natural chitosan nanofibers. Colloids Surfaces B Biointerfaces. 2011;**84**(1):155-162

[53] Iwamoto S, Nakagaito AN, Yano H, Nogi M. Optically transparent composites reinforced with plant fiber-based nanofibers. Applied Physics A: Materials Science & Processing. 2005;**81**(6):1109-1112

[54] Meneghetti P, Qutubuddin S. Synthesis, thermal properties and applications of polymer-clay nanocomposites. Thermochimica Acta. 2006;**442**(1-2):74-77

[55] Zhao M, Forrester J V, McCaig CD. A small, physiological electric field orients cell division. Proceedings of the National Academy of Sciences of the United States of America. 1999;**96**:4942-4946

[56] Yao L, Shanley L, Mccaig C, Zhao M. Small applied electric fields guide migration of hippocampal neurons. Journal of Cellular Physiology. 2008;**216**(2):527-535

[57] Woo DG, Shim MS, Park JS, Yang HN, Lee DR, Park KH. The effect of electrical stimulation on the differentiation of hESCs adhered onto fibronectin-coated gold nanoparticles. Biomaterials. 2009;**30**(29):5631-5638

[58] Sisken BF, Walker J, Orgel M. Prospects on clinical applications of electrical stimulation for nerve regeneration. Journal of Cellular Biochemistry. 1993;**51**(4):404-409

[59] Zhao M. Electrical fields in wound healing-An overriding signal that directs cell migration. Seminars in Cell and Developmental Biology. 2009;**20**(6):674-682

[60] Kirson ED, Dbalý V, Tovaryš F, Vymazal J, Soustiel JF, Itzhaki A, et al. Alternating electric fields arrest cell proliferation in animal tumor models and human brain tumors. Proceedings of the National Academy of Sciences of the United States of America. 2007;**104**(24):10152-10157

[61] Fabbri P, Valentini L, Bittolo Bon S, Foix D, Pasquali L, Montecchi M, et al. In-situ graphene oxide reduction during UV-photopolymerization of graphene oxide/acrylic resins mixtures. Polymer (Guildf). 2012;**53**(26):6039-6044

[62] Huang YL, Tien HW, Ma CC, Yang SY, Wu SY, Liu HY, et al. Effect of extended polymer chains on properties of transparent graphene nanosheets conductive film. Journal of Materials Chemistry. 2011;**21**(45):18236

[63] Huang P, Chen W, Yan L. An inorganic-organic double network hydrogel of graphene and polymer. Nanoscale. 2013;**5**(13):6034-6039

[64] Wan S, Hu H, Peng J, Li Y, Fan Y, Jiang L, et al. Nacre-inspired integrated strong and tough reduced graphene oxide-poly(acrylic acid) nanocomposites. Nanoscale. 2016;**8**(10):56495656

[65] Iijima S. Helical microtubules of graphitic carbon. Nature. 1991;**354**(6348):56-58

[66] Rao CNR, Satishkumar BC, Govindaraj A. Nanotubes. ChemPhysChem. 2001;**2**:78-105

[67] Wang ZG, Wang Y, Xu H, Li G, Xu ZK. Carbon Nanotube-Filled nanofibrous membranes electrospun from Poly(acrylonitrile- co -acrylic acid) for glucose biosensor. The Journal of Physical Chemistry C. 2009;**113**(7):2955-2960

[68] Logakis E, Pandis CH, Pissis P, Pionteck J, Pötschke P. Highly conducting poly(methyl methacrylate)/carbon nanotubes composites: Investigation on their thermal, dynamic-mechanical, electrical and dielectric properties. Composites Science and Technology. 2011;**71**(6):854-862

[69] Huang ZH, Qiu KY. The effects of interactions on the properties of acrylic polymers/silica hybrid materials prepared by the *in situ* sol-gel process. Polymer (Guildf). 1997;**38**(3):521-526

[70] Ramanathan T, Abdala AA, Stankovich S, Dikin DA, Herrera-Alonso M, Piner RD, et al. Functionalized graphene sheets for polymer nanocomposites. Nature Nanotechnology. 2008;**3**(6):327-331

[71] Laachachi A, Cochez M, Ferriol M, Lopez-Cuesta JM, Leroy E. Influence of TiO2 and Fe2O3 fillers on the thermal properties of poly(methyl methacrylate) (PMMA). Materials Letters. 2005;**59**(1):36-39

[72] Laachachi A, Leroy E, Cochez M, Ferriol M, Lopez Cuesta JM. Use of oxide nanoparticles and organoclays to improve thermal stability and fire retardancy of poly(methyl methacrylate). Polymer Degradation and Stability. 2005;**89**(2):344-352

[73] Aymonier C, Bortzmeyer D, Thomann R, Muelhaupt R. Poly (Methyl methacrylate)/ Palladium nanocomposites: Synthesis and characterization of the morphological, thermomechanical, and thermal properties. Chemistry of Materials. 2003;**15**(25):4874-4878

[74] Roorda WE, Bouwstra JA, de Vries MA, Junginger HE. Thermal analysis of water in p(HEMA) hydrogels. Biomaterials. 1988;**9**(6):494-499

[75] Serrano-Aroca Á, Monleón-Pradas M, Gómez-Ribelles JL, Rault J. Thermal analysis of water in reinforced plasma-polymerised poly(2-hydroxyethyl acrylate) hydrogels. European Polymer Journal. 2015;**72**:523-534

[76] Tanaka M, Mochizuki A. Effect of water structure on blood compatibility--thermal analysis of water in poly(meth)acrylate. Journal of Biomedical Materials Research Part A. 2004;**68**(4):684-695

[77] Serrano-Aroca Á, Campillo-Fernández AJ, Gómez-Ribelles JL, Monleón-Pradas M, Gallego-Ferrer G, Pissis P. Porous poly(2-hydroxyethyl acrylate) hydrogels prepared by radical polymerisation with methanol as diluent. Polymer (Guildf). 2004;**45**(26):8949-8955

[78] Clayton AB, Chirila T V, Lou X. Hydrophilic Sponges based on 2-Hydroxyethyl methacrylate. V. Effect of crosslinking agent reactivity on mechanical properties. Polymer International. 1997;**44**:201-207

[79] Monleón-Pradas M, Gómez-Ribelles JL, Serrano-Aroca Á, Gallego-Ferrer G, Suay-Antón J, Pissis P. Interaction between water and polymer chains in poly(hydroxyethyl acrylate) hydrogels. Colloid and Polymer Science. 2001;**279**(4):323-330

[80] Monleón-Pradas M, Gómez-Ribelles JL, Serrano-Aroca Á, Gallego Ferrer G, Suay Antón J, Pissis P. Porous poly (2-hydroxyethyl acrylate) hydrogels. Polymer (Guildf). 2001;**42**(10):4667-4674

[81] Ahmed EM. Hydrogel: Preparation, characterization, and applications: A review. Journal of Advanced Research. 2015;**6**(2):105-121

[82] Tanaka T. Gels. Scientific American. 1981;**244**(1):124-136, 138

[83] Williams MMR. The mathematics of diffusion. Annals of Nuclear Energy. 1977;**4**(4-5): 205-206

[84] Wang X, Zhang Y, Hao C, Dai X, Zhu F, Ge C. Ultrasonic synthesis and properties of a sodium lignosulfonate–grafted poly(acrylic acid-co-acryl amide) composite super absorbent polymer. New Journal of Chemistry. 2014;**38**(12):6057-6063

[85] Gehrke SH, Biren D, Hopkins JJ. Evidence for Fickian water transport in initially glassy poly(2-hydroxyethyl methacrylate). Journal of Biomaterials Science Polymer Edition. 1995;**6**(4):375-390

[86] Wang J, Wu W. Swelling behaviors, tensile properties and thermodynamic studies of water sorption of 2-hydroxyethyl methacrylate/epoxy methacrylate copolymeric hydrogels. European Polymer Journal. 2005;**41**(5):1143-1151

[87] Kavitha T, Kang IK, Park SY. Poly(acrylic acid)-grafted graphene oxide as an intracellular protein carrier. Langmuir. 2014;**30**(1):402-409

[88] Krishna-Rao KSV, Ha CS. PH Sensitive hydrogels based on acryl amides and their swelling and diffusion characteristics with drug delivery behavior. Polymer Bulletin. 2009;**62**(2):167-181

[89] Zhang XZ, Yang YY, Wang FJ, Chung TS. Thermosensitive poly(N-isopropylacrylamide-co-acrylic acid) hydrogels with expanded network structures and improved oscillating swelling-deswelling properties. Langmuir. 2002;**18**(6):2013-2018

[90] Shi L, Chen J, Teng L, Wang L, Zhu G, Liu S, et al. The antibacterial applications of graphene and its derivatives. Small. 2016; **12**(31):4165-4184

[91] He J, Söderling E, Lassila LVJ, Vallittu PK. Incorporation of an antibacterial and radiopaque monomer in to dental resin system. Dental Materials Journal. 2012;**28**(8):e110–e117

[92] Ji H, Sun H, Qu X. Antibacterial applications of graphene-based nanomaterials: Recent achievements and challenges. Advanced Drug Delivery Reviews. 2016;**105**:176-189

[93] Moszner N, Salz U. New developments of polymeric dental composites. Progress in Polymer Science. 2001;**26**(4):535-576

[94] Guo C, Zhou L, Jianxiong L. Effects of expandable graphite and modified ammonium polyphosphate on the flame-retardant and mechanical properties of wood flour-polypropylene composites. Polymer Composites. 2013;**21**(7):449-456

[95] Kong H, Jang J. Antibacterial properties of novel poly(methyl methacrylate) nanofiber containing silver nanoparticles. Langmuir. 2008;**24**(5):2051-2056

[96] González-Sánchez MI, Perni S, Tommasi G, Morris NG, Hawkins K, López-Cabarcos E, et al. Silver nanoparticle based antibacterial methacrylate hydrogels potential for bone graft applications. Materials Science and Engineering: C. 2015;**50**:332-340

[97] Fan Z, Liu B, Wang J, Zhang S, Lin Q, Gong P, et al. A novel wound dressing based on Ag/graphene polymer hydrogel: Effectively kill bacteria and accelerate wound healing. Advanced Functional Materials. 2014;**24**(25):3933-3943

[98] Ping X, Wang M, Xuewu G. Surface modification of poly(ethylene terephthalate) (PET) film by gamma-ray induced grafting of poly(acrylic acid) and its application in antibacterial hybrid film. Radiation Physics and Chemistry. 2011;**80**(4):567-572

[99] Annabi N, Nichol JW, Zhong X, Ji C, Koshy S, Khademhosseini A, et al. Controlling the porosity and microarchitecture of hydrogels for tissue engineering. Tissue Engineering Part B: Reviews. 2010;**16**(4):371-383

[100] Arora KA, Lesser AJ, McCarthy TJ. Compressive behavior of microcellular polystyrene foams processed in supercritical carbon dioxide. Polymer Engineering & Science. 1998;**38**(12):2055-2062

[101] Thomson RC, Wake MC, Yaszemski MJ, Mikos AG. Biodegradable polymer scaffolds to regenerate organs. Advances in Polymer Science. 1995;**122**:247-277

[102] Andrianova GP, Pakhomov SI. Porous materials from crystallizable polyolefins produced by gel technology. Polymer Engineering & Science. 1997;**37**(8):1367-1380

[103] Serrano-Aroca Á, Monleón-Pradas M, Gómez-Ribelles JL. Macroporous poly(methyl methacrylate) produced by phase separation during polymerisation in solution. Colloid and Polymer Science. 2007;**285**(7):753-760

[104] Flynn L, Dalton PD, Shoichet MS. Fiber templating of poly(2-hydroxyethyl methacrylate) for neural tissue engineering. Biomaterials. 2003;**24**(23):4265-4272

[105] Brígido-Diego R, Pérez-Olmedilla M, Serrano-Aroca Á, Gómez-Ribelles JL, Monleón-Pradas M, Gallego-Ferrer G, et al. Acrylic scaffolds with interconnected spherical pores and controlled hydrophilicity for tissue engineering. Journal of Materials Science: Materials in Medicine. 2005;**40**(18):4881-4887

[106] Kang HW, Tabata Y, Ikada Y. Fabrication of porous gelatin scaffolds for tissue engineering. Biomaterials. 1999;**20**(14):1339-1344

[107] Dragusin DM, Van Vlierberghe S, Dubruel P, Dierick M, Van Hoorebeke L, Declercq HA, et al. Novel gelatin–PHEMA porous scaffolds for tissue engineering applications. Soft Matter. 2012;**8**(37):9589

[108] Agarwal S, Wendorff JH, Greiner A. Use of electrospinning technique for biomedical applications. Polymer (Guildf). 2008;**49**(26):5603-5621

[109] Chia HN, Wu BM. Recent advances in 3D printing of tissue engineering scaffolds. Journal of Biological Engineering. 2015;**9**(4):2-14

[110] Derby B. Printing and prototyping of tissues and scaffolds. Science. 2012;**338**(6109): 921-926

[111] Horák D, Hlídková H, Hradil J, Lapčíková M, Šlouf M. Superporous poly(2-hydroxyethyl methacrylate) based scaffolds: Preparation and characterization. Polymer (Guildf). 2008;**49**(8):2046-2054

[112] Kubinová Š, Horák D, Kozubenko N, Vaněček V, Proks V, Price J, et al. The use of superporous Ac-CGGASIKVAVS-OH-modified PHEMA scaffolds to promote cell adhesion and the differentiation of human fetal neural precursors. Biomaterials. 2010;**31**(23):5966-5975

[113] Velasco D, Benito L, Fernández-Gutiérrez M, San Román J, Elvira C. Preparation in supercritical CO_2 of porous poly(methyl methacrylate)-poly(l-lactic acid) (PMMA-PLA) scaffolds incorporating ibuprofen. The Journal of Supercritical Fluids. 2010;**54**(3):335-341

[114] Barry JJA, Silva MMCG, Cartmell SH, Guldberg RE, Scotchford CA, Howdle SM. Porous methacrylate tissue engineering scaffolds: Using carbon dioxide to control porosity and interconnectivity. Journal of Materials Science. 2006;**41**(13):4197-4204

[115] Park K, Hyun JJ, Kim JJ, Ahn KD, Dong KH, Young MJ. Acrylic acid-grafted hydrophilic electrospun nanofibrous poly(L-lactic acid) scaffold. Macromolecular Research. 2006;**14**(5):552-558

[116] Kubinová Š, Horák D, Syková E. Cholesterol-modified superporous poly(2-hydroxyethyl methacrylate) scaffolds for tissue engineering. Biomaterials. 2009;**30**(27):4601-4609

[117] Altmann B, Steinberg T, Giselbrecht S, Gottwald E, Tomakidi P, Bächle-Haas M, et al. Promotion of osteoblast differentiation in 3D biomaterial micro-chip arrays comprising fibronectin-coated poly(methyl methacrylate) polycarbonate. Biomaterials. 2011;**32**(34):8947-8956

[118] Möller S, Weisser J, Bischoff S, Schnabelrauch M. Dextran and hyaluronan methacrylate based hydrogels as matrices for soft tissue reconstruction. Biomolecular Engineering. 2007;**24**(5):496-504

Section 2

Acrylates in Drug Delivery Systems and Green Emission of Light

Chapter 5

Synthetic Polymer-Based Nanoparticles: Intelligent Drug Delivery Systems

Velichka Andonova

Additional information is available at the end of the chapter

http://dx.doi.org/10.5772/intechopen.69056

Abstract

One of the most promising strategies to improve the bioavailability of active pharmaceutical ingredients is based on the association of the drug with colloidal carriers, for example, polymeric nanoparticles, which are stable in biological environment, protective for encapsulated substances and able to modulate physicochemical characteristics, drug release and biological behaviour. The synthetic polymers possess unique properties due to their chemical structure. Some of them are characterized with mucoadhesiveness; another can facilitate the penetration through mucous layers; or to be stimuli responsive, providing controlled drug release at the target organ, tissues or cells; and all of them are biocompatible and versatile. These are suitable vehicles of nucleic acids, oligonucleotides, DNA, peptides and proteins. This chapter aims to look at the 'hot spots' in the design of synthetic polymer nanoparticles as an intelligent drug delivery system in terms of biopharmaceutical challenges and in relation to the route of their administration: the non-invasive—oral, transdermal, transmucosal (nasal, buccal/sublingual, vaginal, rectal and ocular) and inhalation routes—and the invasive parenteral route.

Keywords: poly(ε-caprolactone), poly(lactide-co-glycolide), Eudragit, carbopol, poly(vinyl alcohol), acrylates, vinyl polymers, methacrylates, drug delivery, route of administration

1. Introduction

There is no uniform definition of drug delivery systems (DDSs). Generally, a drug delivery system consists of one or more drug compounds, the technology which carries out the drug(s) inside of the body (medical device or dosage form) and the drug-release mechanism [1]. According to the European Pharmacopoeia (Ph. Eur. 8.0; pg.777), conventional-release (or immediate-release) dosage forms are preparations showing a release of the active substance(s) which is not deliberately modified by a special formulation design and/or manufacturing method. These forms often suffer from some drawbacks in terms of higher dose of the active pharmaceutical ingredients (APIs), lower effectiveness, toxicity and adverse side effects. Modified-release drug delivery systems (MRDDS) have been developed to overcome the disadvantages of the conventional-release dosage forms. They could provide increased efficacy of the API and decreased toxicity/side effects, controlled and/or site-specific delivery, enhanced convenience, lower healthcare cost and better patient compliance.

Roughly, the drug substances suffer from two major problems: solubility and permeability. These two characteristics are responsible for the drug bioavailability upon oral administration and are the basis used to classify the APIs into four fundamental classes; a methodology known as the Biopharmaceutical Classification System (BCS), launched by Amidon and co-workers [2]. One of the most promising strategies to improve their bioavailability is based on the association of API with colloidal carriers, for example, polymeric nanoparticles (NPs), which are stable in biological environment, protective for encapsulated substances and able to modulate physicochemical characteristics, drug release and biological behaviour. Particular attention has to be paid to NPs made by synthetic polymers. These polymers possess unique properties due to their chemical structure, the type of the functional groups in the molecule, the degree of polymerization, the method of synthesis etc. [3–9]. For example, acrylates are pharmacologically inactive and due to their film-forming characteristics, these possess a good compatibility with mucosal membranes. Some of them are insoluble at physiological pH values and capable of swelling as opposed to the others, pH-responsive polymers, which are soluble only at pH 6–7. Those of them which are polycationic polymers are characterized by better mucoadhesive properties [10]. Furthermore, these are suitable vehicles of nucleic acids, oligonucleotides, DNA, peptides and proteins [11]. Aliphatic polyesters, such as poly(ε-caprolactone) (PCL), poly(lactic acid) (PLA) and their co-polymers, are the most exploited polymers because of their biodegradability, biocompatibility and versatility [12–14].

This chapter aims to look at the 'hot spots' in the design of synthetic polymer NPs as an intelligent drug delivery system in terms of biopharmaceutical challenges and in relation to the route of their administration: the non-invasive—oral, transdermal, transmucosal (nasal, buccal/sublingual, vaginal, rectal and ocular) and inhalation routes—and the invasive parenteral route.

2. The challenge: synthetic-based polymeric drug delivery systems and routes of administration

For the preparation of an optimal therapeutic DDS, the following knowledge is required: (i) the physiological characteristics of the organ and tissues in which the API will be absorbed, (ii) the mechanism of absorption and the degree of loss of the API in the biological layers before absorption, (iii) the influence of the properties of the active substance and the drug formulation on the process of absorption and (iv) the possibility of enhancing the bioavailability in the biological tissues and increasing the therapeutic activity of the API, by using appropriate technological approaches to create a stable, tolerable and effective drug formulation. NPs possess an incredible potential to be the most effective DDS. These can (i) protect the drug from the hazardous environment, (ii) affect its solubility in the biological medium, (iii) improve its permeability through the biological membranes, (iv) provide a target drug delivery, (v) permit the administration of a lower dose, (vi) enhance the drug bioavailability, (vii) reduce the systemic side effects and drug toxicity and (viii) ensure a high patient compliance.

2.1. Nanoparticle drug delivery systems obtained from synthetic polymers for oral administration

Oral route of drug administration is preferred by the patients. Therefore, these dosage forms are always in the researchers' focus. It is well known that if a drug to be absorbed in the body, it should have to be in the soluble state. This is particularly true for the oral drug administration. Let us look at what kind of technological problems can create the APIs from different classes of BCS.

Class I: The drug substances which are characterized with both high solubility and high permeability fall in this class. Usually, they do not have bioavailability problems because they dissolve fast and quantitatively, and are readily taken up by intestine. A very fast increase in blood plasma levels is observed for these APIs. But sometimes, from a pharmacokinetic perspective, slower and longer lasting action would be desirable. Polymer-based nanoparticle formulations that deliberately retard drug dissolution are an option for driving the kinetics in the direction of prolonged-release systems.

Class II: Class II APIs represent the largest class of substances in today's drug delivery pipelines. These drugs easily penetrate the relevant physiological barriers but suffer from poor solubility in the aqueous body fluids. There are three main technological approaches to overcome this issue, such as the usage of surfactants, complex formation and nanotechnology. The application of the last one is being discussed later in this chapter.

Class III: These APIs are soluble in the aqueous body fluids, but suffer from low permeability. Most likely they will be excreted without exercising a physiological effect. The use of selective penetration enhancers is a mechanism that can increase the permeation of molecules through the gastrointestinal wall.

Class IV: The drugs falling in this class suffer from both low solubility and low permeability. However, in such cases, prodrugs with enhanced dissolution and permeability that will be converted into active agents under physiological conditions are certainly a possibility to consider.

In this regard, the usage of nanoparticle-based DDSs can be a very promising approach in two opposite directions: (i) To modulate drug dissolution (Class I) in order to achieve desired release kinetics and prolonged release and (ii) to improve the solubility of poorly water-soluble drugs (Class II) which represent the main group of APIs [15–17]. Synthetic-based polymer NPs may solve these problems for the oral delivering of proteins and vaccines. These are promising candidates as drug- and gene-carriers, and could be an elegant approach to cancer-targeted drug delivery.

On the other hand, another challenge of the oral delivery is a result of the obstacles presented by the gastrointestinal tract (GIT) namely, exposure to a wide range of pH environments, enzymatic degradation and poor permeability across the intestinal epithelium. Polymeric NPs could be a very promising approach to provide enhanced drug stability to overcome the mucus barrier (mucus-penetrating biopolymers) or to interact with the intestinal mucus layer (mucoadhesives) increasing the residence time and contact with the epithelium, and to ensure target drug delivery. As a result, an increased concentration of released APIs at the site of absorption could be achieved [18].

The diabetes mellitus is a socially significant disease which affects many people and will increase to 438 million worldwide by the year 2030, according to the World Health Organization (WHO) [19, 20]. Insulin therapy is the best choice for the clinical management of type I diabetes mellitus but its subcutaneous administration leads to a poor patient compliance. Oral route could be preferable but insulin cannot be well absorbed orally because of its rapid enzymatic degradation in the GIT. Polymer NPs have huge potential for the effective oral delivery of insulin. According to Alai et al. [20], polymeric NPs and micelles could provide better drug stability in the harsh GIT environment and enhanced drug transport ability. It could be ensured by biocompatible polymers with mucoadhesive and absorption-enhancing properties (e.g. Eudragit L100-55 and Eudragit S100, poly(lactide-co-glycolide) (PLGA), PCL, alginate, chitosan and dextrane).

Recently, Gutjahr et al. [21] have highlighted the potential application of biodegradable polymeric NPs-based vaccine adjuvants for lymph nodes targeting. The authors have presented the different PLGA, PLA and PCL-based nanoparticulate adjuvants as innovative systems, capable of co-delivering immunopotentiators and antigens which may (i) enhance the drug delivery, (ii) increase the persistence into lymph nodes and promote a mature immune response and (iii) direct the response to a specific antigen and allow the induction of a cytotoxic immune response. This approach could be very promising to limit the spread of diseases caused by HIV, *Chlamydia trachomatis* and *Bacillus anthracis*.

Ma et al. [22] have developed a novel delivery approach for tumour antigenic peptides in order to elicit enhanced immune responses using PLGA-NPs encapsulating tumour antigenic peptides. They have found that human dendritic cells (DCs) loaded with PLGA-NPs encapsulating a peptide cocktail induced a significantly stronger antigen-specific T lymphocytes response in comparison with those including free peptide. The PLGA-NPs loaded with a 63 times lower peptide dose showed more prominent antigen-specific T lymphocytes response *in vivo* than that emulsified in incomplete Freud's adjuvant.

Biomaterial-based nanoparticulate delivery systems that encapsulate a plasmid DNA represent a better strategy for DNA vaccine delivery compared to those delivered as naked plasmid DNA due to their degradation and inefficiency [23]. The PLA-poly(ethylene glycol) (PLA-PEG) NPs containing a high loading of plasmid DNA in a free form or co-encapsulated with either poly(vinyl alcohol) (PVA) or poly(vinyl pyrrolidone) (PVP) have been prepared by different techniques [24]. The researchers have found that plasmid DNA can be very efficiently encapsulated into PLA-PEG NPs and, depending on the processing conditions, these NPs release plasmid DNA either very rapidly or in a controlled manner. Furthermore, NPs-in-microsphere oral system (NiMOS) for gene delivery and transfection in specific regions of the GIT has been developed and evaluated [25]. Plasmid DNA has been encapsulated in type B gelatin NPs. NiMOS have been prepared by further protecting the DNA-loaded NPs in a PCL matrix to form microspheres. The results of biodistribution studies showed that NiMOS resided in the stomach and small intestine for relatively longer duration. After 5 days of post-oral administration, the authors have observed transgene expression in the small and large intestine of rats. Based on these results, NiMOS could be considered as a potential gene delivery vehicle for therapeutic and vaccination purposes.

Environmentally responsive biomaterials are commonly used to achieve controlled drug release in the GIT [26, 27]. The pH-responsive polymers (e.g. Eudragit), mucoadhesives (e.g. carbopol), enzyme-responsive (e.g. guar gum) and pressure-sensitive polymers (e.g. PEG-ethylcellulose) are frequently incorporated into nanoparticle-loaded microdelivery devices to improve oral drug delivery [28].

Recently, a class of lipid-like materials termed 'lipidoids' (synthesized by the Michael addition of alkyl-amines to alkyl-acrylates) have been shown as potential delivery systems of siRNA to the liver and immune cells [29]. The authors established that a single 10 nM dose of siRNA-loaded lipidoid NPs depressed GAPDH mRNA expression for a week and provided powerful, dose-dependent and stable gene silencing in *Caco-2* cells. Moreover, they found no significant induction of cytotoxicity in cells or changes in intestinal barrier function. In this regard, the potential of lipidoid NPs for the treatment of intestinal disorders can be emphasized.

The investigations presented above are just a few examples that strongly suggest the enormous potential of polymeric-based NPs as DDSs for oral administration directing to improve the bioavailability of APIs.

2.2. State-of-the-art topical and transdermal drug delivery nano-carriers

Skin is the largest organ in the human body and it acts as a permeation barrier, mainly due to the *stratum corneum* which is a part of its structure. On the other hand, this large area could be used as a unique delivery pathway for APIs. They can penetrate (i) into skin strata providing topical drug delivery and (ii) through subcutaneous tissues and pass into systemic circulation providing transdermal delivery. It is well known that only APIs characterized by moderate lipophilicity and molecular weight less than 500 Da are able to permeate the *stratum corneum* and penetrate into deeper layers of the skin. Passive and active permeation enhancement methods have been widely applied to increase the skin penetration. Zhang et al. [30] have presented in depth the permeation enhancement methods as well as the major challenges for the treatment of various dermatological diseases. They have put the focus on the penetration of ultra-small NPs into skin strata, the targeted delivery of the encapsulated APIs to hair follicle stem cells, and the combination of NPs and microneedle array technologies for special applications, such as vaccine delivery. Recent literature has demonstrated that NP-based DDSs for topical application can be very successful due to the chemical and physical protection of drug used, controlled release, and cell and tissue-specific targeting [11, 31–33]. These systems combine the advantages of both the nanosized drug carriers and the topical approach, and are promising for the treatment of various skin diseases providing a high patient compliance.

A few studies have described the successful application of tyrosine-derived nanospheres (ThyroSpheres™) as DDSs of lipophilic molecules like paclitaxel and Vitamin D3 [30, 34, 35]. These polymeric nanospheres provide sustained drug release, improve the skin delivery and enhance the chemical stability of drug used. In another study, Batheja et al. [36] have investigated a gel formulation (carbopol and hydroxypropyl methylcellulose (HPMC)) containing tyrosine-derived nanospheres. The authors have found that dispersion of Nile Red-loaded nanospheres in 1% w/v HPMC gel (i) did not show any short-term cellular toxicity or tissue irritation, (ii) the deposition of Nile Red via the nanosphere gel in the upper and lower dermis has been 1.4- and 1.8-fold higher, respectively, than the amount of Nile Red deposited via an aqueous nanosphere formulation and (iii) Azone (0.2 M) incorporation into nanosphere gel formulation led to a 1.4-fold additional increase of drug deposition in porcine *stratum corneum* and epidermis. In this regard, ThyroSpheres™ dispersed in gels could provide improved topical delivery of lipophilic drugs and agents for personal hygiene.

Pharmacokinetics and anti-inflammatory effect of a novel carbopol 934 gel system containing ketoprofen-methylcellulose solid NPs have been studied by Nagai et al. [37]. The authors have established that the penetration rate (Jc) and penetration coefficient through the skin (Kp) values of the ketoprofen-NP-loaded gel have been significantly higher than those of gel containing ketoprofen micro-particles as well as apparent absorption rate constant (ka), area under the curve (AUC) and the amounts of the drug in the skin of rats have also been significantly higher than those of rats receiving the ketoprofen micro gel. These findings suggest that the topical DDS using NPs could lead to an expansion in the therapeutic use of the drug.

Methyl methacrylate copolymers (Eudragit®) have been exploited to develop transdermal patches, medicated plasters (hereinafter patches), film-forming sprays, microsponges and NPs intended to be applied on the skin. Cilurzo et al. [38] have reviewed the information regarding the application of Eudragits in the design and development of these dosage forms focusing on the impact of formulative variables on the skin drug penetration and the patch adhesive properties. The authors have reported that a strict connection between the matrix hydrophilicity and drug penetration probably exists. Moreover, micro- and nano-systems exploiting the ionizable nature of some Eudragits can offer novel opportunities to develop pH-sensitive DDSs suitable for triggering its release onto the skin.

Transcutaneous immunization is a promising vaccination strategy for the treatment of infectious diseases and cancer. Rancar et al. [39] have studied PLA and polystyrene (PS) particle-based antigen (HIV-1 p24 protein) delivery across partially disrupted skin barrier (cyanoacrylate skin surface stripping). The authors have established that the polymer particles targeted HIV-1 p24 protein to the hair follicles and it has been found in skin cells, especially in Langerhans cells and dermal DCs after diffusion of p24 protein to the epidermis and dermis. The researchers have concluded that particle-based antigen delivery across partially disrupted skin barrier is a feasible and effective approach to needle-free transcutaneous vaccination.

Microneedle skin patches represent an attractive technology for non-invasive transcutaneous delivery of vaccines. These DDSs use the accessibility and proven immune competence of the skin for enhanced immunity. They mimic several aspects of cutaneous pathogen invasion by targeting antigen to skin-resident DCs and triggering local inflammatory responses in the skin, which are correlated with enhanced immune responses. DeMuth et al. [40] have tested whether the control over vaccine delivery kinetics can enhance the immunity through further mimicry of the kinetic profiles presented during natural acute infections. The authors have prepared microneedles which consist of a silk tip and a poly(acrylic acid) (PAA) base. The skin application of microneedle patches to deliver a vaccine with improved release kinetics led to >10-fold increases in antigen-specific T cell and humoral immune responses compared to the traditional parenteral immunization.

2.3. Synthetic-based nano- and micro-particles intended for transmucosal drug delivery

Transmucosal routes of drug delivery include transport across the nasal, rectal, vaginal, ocular and oral cavity mucous membranes, and offer distinct advantages over oral administration for systemic drug delivery. The main advantages of this route include a possible circumvention of first-pass effect and avoidance of pre-systemic elimination within the GIT.

2.3.1. Intranasal nanoparticle-based drug delivery systems

Different approaches to improve the nasal drug bioavailability have been described in the scientific literature. The strategies are limited to (i) enhancing the nasal absorption, (ii) modifying the structure of the drug and the physicochemical properties and (iii) increasing the residence time. Polymeric nano- and micro-particles can reduce the processes of drug degradation by chemical derivatization or covalent bonding and increase the residence time on the mucous membranes. These DDSs may include in addition enzyme inhibitors, promoters of absorption and/or mucoadhesive polymers in order to increase stability, membrane permeability and retention time in the nasal cavity. The intranasal route is successfully used for drug delivery of low molecular weight APIs with non-peptide structure, peptides (insulin, calcitonin and thyroid hormones), vaccines and direct delivery in the central nervous system.

Intranasal delivery seems to be a promising approach for drug delivery across the blood-brain barrier (BBB) to the brain, providing a significant advantage over currently used strategies without damaging the BBB. Alzheimer's disease is a socially significant neurological disorder that results in cognitive and behavioural impairment. It affects many people all over the world and their number increases rapidly. Fonseca-Santos et al. [41] have presented an extended review about the significant benefits that intranasal polymer nanoparticle-based DDSs could ensure in the treatment of this disorder. Polymeric NPs (quinoline-n-butyl cyanoacrylate-based NPs, rivastigmine-loaded poly(n-butyl cyanoacrylate NPs) coated with polysorbate 80 and polysorbate 80-coated solid lipid NPs have been used for both diagnostic and treatment of Alzheimer's disease.

Although the results presented by Zhuang et al. [42] suggest that intranasal delivery of an anti-inflammatory agent, such as curcumin, and the anti-Stat3 agent, JSI-124, provides a promising non-invasive approach for the treatment of brain inflammatory-related diseases, such as malignant gliomas, biosafety considerations have been challenging. Recently, the researchers have developed a grapefruit-derived nanovector hybrid with polyethylenimine (pGNV) for effective intranasal delivery of miRNA to the brain [43]. The authors have found that the hybrid not only enhanced the capacity to carry RNA but also eliminated the toxicity of the polyethylenimine. Enhanced targeting has been further achieved by coating pGNVs with the tumour targeting moiety, folic acid.

The use of polymeric carriers for drug delivery to the brain via the nose-to-brain route holds great promise, on the basis of pre-clinical research and clinical data. On the other hand, a strict toxicity assessment of NPs regarding to the morphology and functions of nasal mucosa, the target drug delivery and the biopharmaceutical characteristics and pharmacokinetics of NPs is needed before these find a clinical utility [44]. The regulatory agencies recommend the implementation of 'Quality by Design' (QbD) and the process optimization for the product development to produce a safe and effective intranasal DDS. The characteristics of the NPs, such as particle size, size distribution, particle shape, surface chemistry and structure are crucial for nanoparticle uptake by the nasal mucosa and will determine the therapeutic effect and possible toxicity.

The PLGA has been widely explored for preparation of polymeric NPs and is well reported for its mucoadhesive properties, improved drug stability and enhanced entrapment efficiencies. Lorazepam-loaded PLGA-NPs have been formulated using a nanoprecipitation approach [45]. This DDS showed controlled lorazepam release and potential outcome which have been optimized using 4-factor, 2-level Box-Behnken design.

Furthermore, stimuli responsive polymers have been widely exploited as nasal DDSs. These smart polymers possess liquid state at room temperature and in response to the nasal temperature, pH and ions present in mucous, can undergo *in situ* gelation in the nasal cavity. These are able not only to enhance the drug retention in the nasal cavity but also to provide controlled release, ease of administration, enhanced drug permeation and protection of the drug from mucosal enzymes. Some of the aspects of the stimuli responsive polymers and their gelling mechanisms have already been discussed [46]. Thermoresponsive polymers (e.g. poloxamer 407) and combination thereof with mucoadhesive carbomers, chitosan and cellulose derivatives are widely used to improve the drug bioavailability. For example, to improve the intranasal absorption of plasmid DNA, Park et al. [47] have designed delivery systems composed of *in situ* gelling poloxamers and mucoadhesive polycarbophil or polyethylene oxide (PEO) polymers. The authors have found that at 3 h post-dose, the nasal tissue levels of plasmid DNA given in poloxamer/polycarbophil and poloxamer/PEO 0.8% have been 10- and 40-fold higher relative to saline. These findings have indicated the safety and effective utilization of *in situ* gelling and mucoadhesive polymers for intranasal plasmid DNA delivery.

In another study, Nakamura et al. [48] have formulated mucoadhesive pH-sensitive budesonide micro-particles of poly(methacrylic acid) and PEG for nasal delivery. Following nasal administration of the budenoside-loaded polymeric micro-particles, the peak plasma concentration has been reached in about 45 min, and the concentration in plasma remained constant for a minimum of 8 h compared to intravenous drug administration where the plasma concentration peaked immediately and decreased rapidly over the next 4 h. Thus, intranasally administered budesonide-polymer DDS possesses enhanced durability of the drug concentration in plasma. Furthermore, polyvinyl acetal dimethyl aminoacetate pH-sensitive gel has provided controlled release of chlorpheniramine maleate and tetrahydrozoline hydrochloride incorporated [49].

All the experimental results presented above disclose the huge potential of the intranasal route for drug administration especially using polymeric-based nano- and micro-particle DDSs. The considerations according to their biosafety, the quantity of API administered nasally that will be transported directly from nose to the target tissues and the mechanism of this transport should be estimated.

2.3.2. *Rectal and vaginal route of drug delivery*

Towards the development of vaginal DDSs, the most often used polymers are PCL, PLA, PLGA, poly(methyl methacrylate), PS etc. They may incorporate low molecular weight APIs as well as nucleic acids for the prevention of viral infections, responsible for genital herpes, AIDS and cervical carcinoma. Usually, these polymers possess good mucoadhesiveness and thereby increasing the residence time, they provide a higher bioavailability. The efforts are aimed at creating muco-penetrating NPs for vaginal and rectal administration [50]. For example, different surface-engineered PCL NPs have been designed to modulate the permeability and retention of dapivirine (microbicide against HIV/AIDS) in vaginal and rectal mucosa [51]. The results presented demonstrated that PEO-modified PCL NPs are very suitable carriers for vaginal and rectal delivery of microbicides due to their ability to modify drug permeability and retention in mucosal tissues.

An even greater challenge is the use of so-called 'gene silencing' in the treatment of vaginal infections and carcinomas. Woodrow et al. [50, 52] have provided significant evidence that siRNA complexes could be successfully delivered by PLGA-NPs, providing sustained gene silencing in the female reproductive tract. Later, Steinbach et al. [50, 53] have showed that siRNA delivery via PLGA-NPs could provide protection against vaginal infection. Furthermore, these results are further evidence that siRNA-loaded PLGA-NPs may provide vaginal protection from sexually transmitted infections with improved safety compared to conventional siRNA delivery vehicles.

Nanofibres have various applications, one of which is drug delivery, especially in local chemotherapy. Recently, drug-loaded ultrafine fibres have been used in local chemotherapy of cervical cancers. Biodegradable PLA fibre mats loaded with paclitaxel showed strong inhibition of xenograft U14 cervical cancer [54]. In another study, *in vivo* trials of cisplatin-loaded PEO/PLA composite electrospun nanofibres demonstrated enhanced anti-tumour efficacy with better systemic safety than the intravenous injection group [55]. This indicates the benefits of localized delivery over systemic delivery. Ordikhani et al. [56] have summarized some of the recent research in systemic and localized DDSs and compared the advantages and disadvantages of these methods in the treatment of cervical cancer.

Yoo et al. [57] have reported the development of pH-responsive NPs prepared from Eudragit® S-100 that are characterized by low encapsulation efficiency for hydrophilic compounds (26%) compared to hydrophobic compounds (71%). Burst release occurred at pH 7.4 in the range expected when semen contacts vaginal mucus because this formulation has been made of pH-sensitive polymer. The pH-sensitive NPs would be a promising carrier for the vaginal-specific delivery of various therapeutic drugs including microbicides and peptides or proteins.

For inflammatory bowel disease treatment, local delivery of molecules loaded in NPs to the inflamed colon could be a promising strategy. Mucoadhesive and pH-sensitive ovalbumin (OVA)-loaded NPs as well as NPs for sustained drug delivery have been obtained from trimethylchitosan (TMC), Eudragit® S100 and a polymer mixture (PLGA, PEG-PLGA and PEG-PCL), respectively, for a target colon delivery to the inflamed tissues [58]. Mannose or a specific peptide has been grafted on the PEGylated NPs to ensure the target drug delivery. The TMC NPs had the highest apparent permeability for OVA in the untreated model. However, in the inflamed model, there was no difference between TMC, PLGA-based and Eudragit® NPs. Mannose-grafted PLGA-NPs showed the highest accumulation of OVA in inflamed colon. Based on these results, active targeting of macrophages and DCs may be a promising approach for targeting the colon in inflammatory bowel disease.

2.3.3. Recent advances in ophthalmic nanoparticulate drug delivery carriers

Prospects for the application of polymeric micro- and nano-carriers as DDSs in ophthalmic preparations are related to an improved solubility of less soluble drugs, a controlled release, a targeted transport and an enhanced chemical stability in order to increase efficiency and reduce side effects, overcoming physiological barriers and delivery of APIs to the posterior segment of the eye, wherein the penetration is usually hampered [59–67]. An important feature of ophthalmic DDSs is the ability of retention in the ocular tissues. In this regard, the characteristics of polymers like mucoadhesivness and option for modelling of their surface properties are crucial. Synthetic-based polymer NPs obtained by polyacrylates, polymethylmetacrylates, polyalkylcyanoacrylates and polyvinyl acetates (PVAc), PCL, PLA, PLGA etc. meet these basic features and their properties as DDSs can be modified [59–66]. Giannavola et al. [67] have found that both, uncoated and PEG-coated acyclovir-loaded PLA NPs, characterized with sustained drug release have been well tolerated, but PEG-coated NPs have shown greater efficacy compared to uncoated NPs. A great number of studies, related to the preparation of nanoparticulate DDSs that provide a good retention and sustained drug release, can be found in the scientific literature. There are examples like indomethacin-loaded PVAc/carbopol NPs, indomethacin-loaded PVAc/chitosan NPs, pilocarpine-loaded chitosan/carbopol NPs, rapamycin-loaded chitosan/PLA NPs, gatifloxacin/prednisolone-loaded NPs of Eudragit RS100 and RL100 coated with hyaluronic acid, sparfloxacin- and levofloxacin-loaded PLGA-NPs, etc. [61, 62, 68–72]. Generally, the authors have concluded that the use of this kind of polymers leads to increased pre-corneal residence time and improve the drug penetration across the cornea. As previously mentioned, these nanoparticulate DDSs provide enhanced drug stability, increased residence time, better bioavailability in ophthalmic tissues, controlled release and good biotolerability.

In the past few years, a variety of novel stimuli responsive ophthalmic DDSs have been reported. The combination of NPs and *in situ* gel has been developed [73, 74]. It is known as 'nanoparticle laden *in situ* gel'. In an extensive review, Kumar et al. have described every aspect of this novel formulation [75]. The polymeric nanoparticle-loaded *in situ* gel provides a sustained and prolonged release. Biodegradable and water-soluble polymers make them more acceptable and excellent DDSs. These *in situ* activated gel-forming systems seem to be favoured as they can be administered in a drop form and produce considerably less blurred vision. Owing to its control of drug release, the dosage form is more acceptable by the patients and thus increases the patient compliance [75].

The bioavailability of ophthalmic drugs can be improved by soft contact lenses-based ophthalmic DDSs. For example, the cross-linked NPs based on PCL, 2-hydroxyethyl methacrylate (HEMA) and poly-ethylene glycol diacrylate (PEG-DA) have been prepared by surfactant-free mini emulsion polymerization. The lens material has been prepared through photopolymerization of HEMA and N-vinylpyrrolidone (NVP) using PEG-DA as cross-linker. NPs and hydrogel showed high viability, indicating the absence of cytotoxicity and stimulatory effect. The drug-release studies revealed that the hydrogel embedded with NPs released the loteprednol etabonate for a period of 12 days [76].

Many polymeric systems have been used to fabricate ocular inserts to improve ocular bioavailability and drug retention. The inserts have shown some advantages like reduced dosing frequency and increased corneal residence time. For example, a cross-linked and Eudragit RL-100-coated ocular insert of gatifloxacin provides better *in vitro* drug release and sustained up to 11 h [77]. Recently, Thakur et al. [78] have prepared bioerodable insert of azithromycin in order to prolong the release time and improve the ocular availability in ophthalmic infections. The model comprising of 1.5% HPMC and 3% Eudragit RL100 has been found to be optimized formulation on the basis of uniformity of thickness and weight, surface pH, folding endurance, percentage moisture loss, percentage moisture absorption, drug content, *in vitro* release, AUC for *in vitro* and *in vivo* release which have been higher than pure drug and shelf life. Furthermore, better ocular tolerability has been found.

GrayBug's controlled release technologies are based on proprietary biodegradable drug-loaded PLGA-NPs, micro-particles and injectable implants providing extended release of small to large molecules for intraocular applications to treat neovascular diseases, such as age-related macular degeneration (AMD), diabetic retinopathy and glaucoma [79].

The Particle Replication in Non-wetting Templates (PRINT) technology offers a unique ability to reproducibly fabricate particles of virtually any size, shape, chemistry, surface functionality, modulus and porosity. Additionally, PRINT has been shown previously to be broadly compatible with a wide range of biodegradable polymers (e.g. PLGA) and molecular entities including small molecules, nucleic acids, enzymes and therapeutic monoclonal antibodies [80]. The unique flexibility of PRINT has been used to develop biodegradable nano- and micro-particle suspensions and biodegradable implants for extended drug delivery into the eye. Such products are ENV515 intra-cameral extended-release prostaglandin analogues and ENV705 extended-release anti-VEGF formulation for AMD therapy. In the ENV705 implant, a trehalose/bevacizumab mixture is dispersed within a polyglycolic acid matrix allowing drug release of effective concentrations over 3–6 months [81].

The use of NPs in ophthalmic formulations can solve most of the problems of drug delivery, mainly related with low bioavailability in the target eye tissues. Mucoadhesive, mucus-penetrating NPs or nanoparticle-loaded *in situ* gelling systems may significantly increase the pre-corneal residence time and ocular penetration, thus improving the drug bioavailability.

The undeniable advantages that these systems provide, as a sustained drug release, a reduced administration frequency and a higher patient compliance, give us grounds to believe that in the next few years, some of them will find their place on the ophthalmic market [82].

2.3.4. Buccal/sublingual route of drug administration

The buccal cavity has a very limited surface area (around 50 cm^2) but the easy access to the site makes it a preferred location for delivering APIs. Buccal route of drug administration is used for the treatment of local diseases of the oral cavity as well as for achievement of systemic effect by avoiding hepatic first-pass metabolism. The sublingual mucosa is relatively more permeable than the buccal mucosa due to the presence of a large number of smooth muscle and immobile mucosa. Therefore, sublingual formulations are designed to release the APIs quickly. The buccal cavity is more suitable for mucoadhesive DDSs and the API could be released in a controlled manner. Bioadhesive micro- and NPs offer more advantages compared to conventional buccal tablets due to their high surface area which allows them to make contact with a larger mucosal surface. The polymers used to prepare these systems must meet the following requirements, such as rapid attachment to the mucosal surface, maintaining a strong interaction which prevents any displacement and the bioadhesion performance should not be impacted by surrounding environmental pH. The various mucoadhesive polymers used for the development of buccal DDSs include cyanoacrylates, PAA, sodium carboxymethylcellulose, hyaluronic acid, hydroxypropyl cellulose, polycarbophil, chitosan and gellan [83].

For example, nanofibre-based mucoadhesive films were invented for oromucosal administration of drug- and vaccines-loaded nano-carriers [84]. The mucoadhesive film consists of (i) an electrospun nanofibrous reservoir layer where the NPs can be reversibly adsorbed or they can be deposited in the pores between the nanofibres, (ii) a mucoadhesive film layer and (iii) a protective backing layer. After mucosal application, nanofibrous reservoir layers are intended to provide prolonged release of NPs into the submucosal tissue. To prove this concept, trans-/intra-mucosal and lymph-node delivery of PLGA-PEG NPs has been demonstrated in a porcine model. This system can mainly be used for sublingual immunization and the development of 'printed vaccine technology'.

Carvedilol nanosuspension has been loaded into mucoadhesive buccal films containing three similar layers: mucoadhesive layer, nanosuspension-containing layer and backing membrane [85]. Carvedilol-loaded nanosuspension has been prepared by a precipitation-ultrasonication method with PVA. Nanosuspension incorporated drug-gel layer was optimized to contain 3% HPMC and 50 mg carbopol 934P. The authors have suggested that the increased relative bioavailability of the obtained formulations was due to the increased surface area of carvedilol and by-passing the hepatic metabolism.

Sapre and Parikh have formulated polymeric NPs-based mucoadhesive system intended for oral mucosal delivery of fluoxetine hydrochloride [86]. In this study, the drug has been encapsulated into poly(methyl vinyl ether/maleic anhydride) (Gantrez MS-955) mucoadhesive NPs. This buccal mucoadhesive system comprising a fluoxetin-loaded NPs layer and an ethyl cellulose layer has been characterized by by-pass first-pass effect, with relatively rapid onset, higher absorption and sustained release effect to increase bioavailability compared to oral absorption.

In another study, Al-Dhubiab has designed and evaluated zolpidem-loaded PLGA nanospheres-impregnated buccal films to prolong the duration of its action [87]. Zolpidem nanospheres have been loaded into mucoadhesive films composed of different concentrations of HPMC K100, Eudragit® RL100 and carbopol 974P. The prepared films showed adequate mucoadhesive strength and excellent physicomechanical strength. The results of the *in vitro* drug-release tests have depicted the potential of the films to provide extended drug release, while *ex vivo* studies have justified the potential of the nanospheres to permeate across the buccal membranes at a controlled rate. Furthermore, *in vivo* the studies have reinforced findings from the *in vitro* and *ex vivo* studies, demonstrating prolonged release and enhanced bioavailability of zolpidem.

Mucoadhesive drug delivery will play an even more important role in delivering of a large number of molecules: new drug molecules due to drug discovery and well-known APIs, which suffer from low solubility, poor bioavailability or chemical instability.

2.4. Nanoparticle-mediated pulmonary drug delivery

The pulmonary route, as a non-invasive method of drug administration for both local and systemic delivery of APIs, is preferable for APIs acting on pulmonary diseases and disorders. Additionally, this route offers many advantages, such as high surface area with rapid absorption due to high vascularization and circumvention of the first-pass effect [88]. The challenges for the pulmonary drug delivery are related with three main clearance mechanisms namely (i) pulmonary clearance, (ii) enzymatic degradation and (iii) rapid systemic absorption. As a result, the inhaled drugs exhibit low bioavailability in the lungs. It is well known that the particulate-based DDSs could solve the problem with drug bioavailability providing (i) drug protection from enzymatic degradation, (ii) evade pulmonary clearance, (iii) target drug delivery to the desired site at the lungs, (iv) controlled drug release, (v) reduce dose frequency, (vi) maximize the therapeutic efficiency and (vii) minimize side effects [89].

For therapeutic purposes, the most commonly used synthetic polymers include PLA, PLGA and PCL. These polymers have numerous advantages mentioned above. Furthermore, several factors, such as aerodynamic diameters, shape and surface properties of these polymer carriers can be tailored and optimized to obtain a particulate-based DDS with high therapeutic efficiency. In an extensive review, El-Sherbiny et al. [89] have presented the factors influencing pulmonary drug deposition and bioavailability as well as the significance of particulate-based pulmonary drug delivery.

For example, rifampicin-loaded PLGA microspheres with adequate aerodynamic properties for lung delivery as aerosols have been recently formulated and studied *in vitro* [90]. The solvent evaporation method with premix membrane homogenization has been applied, with class-3 ethyl acetate as organic solvent, to produce narrowly size-distributed rifampicin-loaded PLGA microspheres for sustained lung delivery as aerosol.

Recently, the development and *in vitro* characterization of PLGA microspheres loaded with totarol (an antibacterial natural drug) for the treatment of long-term bacterial infections has been presented [91]. Moreover, pitavastatin-loaded PLGA-NPs have been designed for the treatment of pulmonary artery hypertension [92]. The authors observed delivery of NPs into alveolar macrophages and small pulmonary arteries for up to 14 days after a single intratracheal administration. The PLGA nanoparticulate-mediated drug delivery has been more effective than systemic administration of pitavastatin, attenuating the development of pulmonary artery hypertension. In addition, treatment with pitavastatin-NPs 3 weeks after monocrotaline injection induced regression of pulmonary artery hypertension and improved survival rate.

The NPs possess sizes that allow them to be easily inhaled and to reach the deep lung. On the other hand, according to the same reason, these could be exhaled. The ideal particle sizes for the pulmonary alveoli administration are between 2 and 5 μm. The Trojan micro-particles contain drug-loaded NPs. These systems offer a compromise between the range of NPs with their main advantages and the benefits which provide micro-particles. Anton et al. [93] have presented the physical principles and experimental procedure involved in the fabrication of these unique systems and their impact on drug delivery and release kinetics. The authors have paid particular attention to the biopharmaceutical application of the Trojan micro-particles. For example, an efficient Trojan delivery of tetrandrine by PVP-block-PCL NPs has shown enhanced apoptotic induction of lung cancer cells and inhibition of its migration and invasion [94].

Furthermore, Simultaneously Manufactured Nano-In-Micro (SIMANIM) particles for the pulmonary delivery of antibodies have been prepared by the spray-drying of a double-emulsion containing human IgG (as a model antibody), lactose, PLGA and dipalmitoylphosphatidylcholine [95]. The continuous release of the model antibody has been observed for 35 days in pH 2.5 release media and released antibody has been shown to be stable and active. 'SIMANIM' particles could be beneficial for the delivery of antibodies targeted against inhaled pathogens or other extracellular antigens, as well as having potential applications in the delivery of a wide range of other APIs.

Nanoparticulate-mediated pulmonary drug delivery provides many advantages compared to the conventional-release dosage forms. Although the synthetic polymer carriers used for pulmonary drug delivery are usually biocompatible and biodegradable, a strict and precise assessment of the potential toxicity and side effects is necessary to be done due to the high surface area and vascularization of the lung.

2.5. Advances in novel parenteral drug delivery systems

The parenteral route of administration is the most effective route for the delivery of the APIs with narrow therapeutic index and poor bioavailability. Major progress has been done in the field of formulation technologies using innovative polymer-based DDSs so as to provide a targeted and sustained release of drug in predictable manner and to overcome the problems associated with conventional parenteral DDSs.

For example, one of the most intriguing strategies to overcome the limitation of classical cytotoxic drugs is their formulation into nanopharmaceutical platforms, that is, nano-carriers, such as liposomes, polymeric NPs and more recently into nanocontainers based on host-guest interactions. Furthermore, biodegradable and injectable *in situ* forming DDSs represent an attractive alternative to microspheres and implants as parenteral depot systems.

Recently, Bao et al. have evaluated daunorubicin (DNR)-loaded PLGA-poly-l-lysine (PLL)-PEG-transferrin (Tf) NPs as a DDS providing sustained release at the specific site and reduced toxicity in normal tissues [96]. The authors have observed a higher drug intra-cellular concentration in *K562* cells, an enhanced anti-cancer activity and a regulation of the expression of some proteins. In another study, to increase the encapsulation of DNR and multi-drug resistance reversal agent tetrandrine (Tet) in the DDS of NPs, Liu et al. [97] have synthesized a functional copolymer PLGA-PLL-PEG, and then it was loaded with DNR and Tet simultaneously (DNR/Tet-PLGA-PLL-PEG-NPs). These NPs have been further modified with transferrin (Tf) due to its specific binding to Tf receptors, which is highly expressed on the surface of tumour cells. The results showed that the accumulated release of DNR and Tet could be sustained over 1 week. Furthermore, the authors have found that the new DNR-loaded NPs have been more effective than DNR alone, inhibiting the cell proliferation of *K562* and *ADR* lines in a dose-dependent manner. The experimental results presented show that PLGA-PLL-PEG-Tf formulation could be very promising DDS having excellent features for target delivery.

7-Ethyl-10-hydroxy camptothecin (SN38) is a potent topoisomerase inhibitor and a metabolite of irinotecan with poor solubility which hampered its clinical development. To overcome this problem, methoxy PEG-2000 (mPEG2K)-SN38 and mPEG2K-PLA1.5K-SN38 conjugates have been prepared and then dispersed into an aqueous medium to form micelles [98]. The authors have found that SN38-loaded micelles with PLA induced a significant tumour inhibition after 30 days compared to those without PLA.

Furthermore, recent report has presented the formulation and evaluation of a core-shell type star polymer with a branched hydrophobic PS core and covalently attached poly(tert-butyl acrylate) arms, as a DDS for cisplatin [99]. The stars were loaded with cisplatin via ligand exchange reaction achieving remarkable high drug payload of 45% (w/w). The release profile of the platinum (II) complexes indicated sustained manner of drug release with no initial burst effect. *In vitro* cell viability study, using different human tumour cell lines, proved that the conjugates exhibited lower cytotoxicity compared to the free agent. Further, the design strategy has been based on functionalization of the polyacrylate arms by a PEGylated cisplatin analogue, allowing for detachment of the coating following hydrolysis in biological environment [100].

It has been noted before that nanoparticulate-based DDSs possess distinct advantages for brain drug delivery. Penetratin (a cell-penetrating peptide with relatively low content of basic amino acids) has been functionalized to PEG-PLA NPs to achieve desirable pharmacokinetic and biodistribution profiles for brain drug delivery [101]. *In vivo* pharmacokinetic and biodistribution studies showed that penetratin-functionalized PEG-PLGA-NPs exhibited a significantly enhanced brain uptake and reduced accumulation in the non-target tissues compared with low-molecular-weight protamine (a cell-penetrating peptide with high arginine content)-functionalized NPs. The application of these penetratin-functionalized NPs can be a promising strategy for brain-targeting drug delivery as well as a basis for the optimization of brain DDSs via surface charge modulation.

Recently, Shamma et al. [102] have investigated injectable *in situ* forming scaffolds loaded with risedronate (bone resorption inhibitor) and with lornoxicam (anti-inflammatory drug) for non-surgical treatment of periapical lesions. They have tested two insoluble copolymers, such as PLGA (ester-terminal) and PLGA-A (acid-terminal). Additionally, sucrose acetate isobutyrate (SAIB) has also been added as a high viscosity water-insoluble carrier as well as porogenic agents like hydrolysed collagen. The scaffolds prepared using 30% (w/v) PLGA or combined PLGA: SAIB (1:1, w/w) with total polymer concentration of 30% (w/v) possessed the most sustained drug-release profile and their application improved the inflammation and enhanced the formation of new bony regions. These results confirm the success of the prepared scaffolds as an innovative approach in the treatment of bone defects.

Nagarajan et al. [103] have synthesized two star-shaped PLA polymers with dipyridamole molecular core as a coating material for making coronary stents. The authors have used L-lactide and DL-lactide for their preparation. The difference in the features of both coating biomaterials was explained with the crystallinity of the polymers synthesized as well as with the fact that the L-form is normally available in humans. The experimental results showed that the new star-shaped PLA polymers with dipyridamol core were bio- and hemo-compatible, and possessed enhanced angiogenic properties. This is another proof for the great importance of the synthetic-based polymer carriers and their application in the pharmaceutical practice.

3. Conclusion and future prospects

The main application of NPs as DDSs and the challenges, regarding to this, lies on the efficient administration of these carriers. Along with all the benefits that these DDSs provide, the choice of route of administration is essential to their performance. The difficulties are related to their absorption at the proper organs, tissues and cells.

Unconditionally, the use of NPs as DDSs may solve most of the problems of drug delivery, mainly related with low bioavailability in the target tissues. Mucoadhesive, mucus-penetrating NPs or 'nano-in-micro-particle' DDSs may significantly increase the retention time on the mucous surface by enhancing the drug absorption and thereby improving drug bioavailability. Nanotechnologies are able to overcome the physiological barriers of the different tissues regarding the route of administration. Polymeric carriers have a stabilizing role on the drug included. They protect it from the unfavourable impact of the environment or the biological fluids. The synthetic-based polymer nano-carriers can assure a controlled release in the target tissue and minimize the side effects. The advantages that these could provide unconditionally improve the drug bioavailability, reduce the administration frequency and dose and enhance the patient compliance. Still not well assessed, the problem with the eventual toxicity and adverse effects of these drug carriers must draw our attention. Although polymeric particles may be biodegradable, their degradation rate must be analysed and toxicity profiles must be assessed in various *in vitro, ex vivo* and *in vivo* models. The development and application of strict safety rules are needed to create an effective and safe drug formulation.

In the future, the main emphasis of investigations will be put on the achievement of a non-invasive drug administration, aiming targeted and controlled release of API with a minimal effective dose. The comprehensive exploration of the problems, associated with the route of drug administration within its complexity, the tissues under normal and pathological conditions, and the multi-compartment pharmacokinetics, will significantly accelerate the further progress in this field.

Author details

Velichka Andonova

Address all correspondence to: andonova_v@abv.bg

Faculty of Pharmacy, Department of Pharmaceutical Sciences, Medical University of Plovdiv, Plovdiv, Bulgaria

References

[1] Tiwari G, Tiwari R, Sriwastawa B, Bhati L, Pandey S, Pandey P, Bannerjee SK. Mint: Drug delivery systems: An updated review. International Journal of Pharmaceutical Investigation. 2012;**2**:2-11. DOI: 10.4103/2230-973X.96920

[2] Amidon GL, Lennernas H, Shah VP, Crison JR. Mint: A theoretical basis for a biopharmaceutic drug classification: The correlation of *in vitro* drug product dissolution and *in vivo* bioavailability. Pharmaceutical Research. 1995;**12**:413-420

[3] Rao JP, Geckeler KE. Mint: Polymer nanoparticles: Preparation techniques and size-control parameters. Progress in Polymer Science. 2011;**36**(7):887-913. DOI: http://dx.doi.org/10.1016/j.progpolymsci.2011.01.001

[4] Arun A, Reddy BSR. Mint: Polymeric drug for antimicrobial activity studies: Synthesis and characterization. Journal of Bioactive and Compatible Polymers. 2003;**18**:219-228

[5] Baskar G, Chandrasekar K, Reddy BSR. Mint: Associated structures of aqueous solution of comblike polymers from 2-acrylamido-2-methyl-1-propanesulfonic acid, dodecylmethacrylate and poly(ethyleneglycol) acrylate macromonomer. Polymer. 2004;**45**:6507-6517

[6] Gaspar LJM, Baskar G, Reddy BSR, Ranganathan R, Peric M. Mint: Effect of N-Glycinyl Maleamic Acid on microstructural characteristics and solubilization properties of SDS micellar assemblies. Langmuir. 2004;**20**:9029-9035

[7] Rani SU, Pandian K, Reddy BSR. Mint: Syntheses and characterization of silver nanoparticles in the acrylate copolymers. Journal of Experimental Nanoscience. 2009;**4**:285-299

[8] Madhavan K, Reddy BSR. Mint: Smaller molecules doing bigger things: A challenge to nanotechnology in separation science. Science Reporter. 2007;**44**(7):17. Available from: www.niscair.res.in/sciencecommunication/popularization%20of%20science/SciRep/scirep2k7/Scirep_sept07.asp [Accessed: 4 February 2017]

[9] Vijay R, Angayarkanny S, Reddy BSR, Mandal AB, Baskar G. Mint: Adsorption and emulsification properties of amphiphilic poly(styrene-co-octadecyl maleamic acid salt) with comb-like architecture. Journal of Colloid and Interface Science. 2010;**346**:143-152

[10] Marschütz MK, Bernkop-Schnürch A. Mint: Thiolated polymers: Self-crosslinking properties of thiolated 450 kDa poly(acrylic acid) and their influence on mucoadhesion. European Journal of Pharmaceutical Sciences. 2002;**15**(4):387-394

[11] Srividhya M, Preethi S, Gnanamani A, Reddy BSR. Mint: Sustained release of protein from poly (ethylene glycol) incorporated amphiphilic comb like polymers. International Journal of Pharmaceutics. 2006;**326**(1-2):119-127

[12] Nagarajan S, Reddy BSR. Mint: Bio-absorbable polymers in implantation-An overview. Journal of Scientific & Industrial Research. 2009;**68**(12):993-1009

[13] Nagarajan S, Reddy BSR. Mint: Preparation of PLA-PDMS-HMDI based polyurethane membranes for drug permeation studies. Journal of Polymeric Materials. 2011;**28**:543-549

[14] Nagarajan S, Reddy BSR, Tsibouklis J. Mint: Fluorescence-detectable, star-shaped polylactic acid construction for implantation. European Polymer Journal. 2012;**48**:1357-1360

[15] Bakrudeen B, Tsibouklis J, Reddy BSR. Mint: Facile fabrication of mesoporous ZnO nanospheres for the controlled delivery of captopril. Journal of Nanoparticle Research. 2013;**15**(3):1505-1509

[16] Lakshmi MS, Sriranjani M, Bakrudeen HB, Kannan AS, Mandal AB, Reddy BSR. Mint: Carvedilol/montmorillonite drug nanoparticles: Processing, characterization and dissolution studies. Applied Clay Science. 2010;**48**:589-593

[17] Madurai SL, Joseph SW, Mandal AB, Tsibouklis J, Reddy BS. Mint: Intestine-Specific, oral delivery of Captopril/Montmorillonite: Formulation and release kinetics. Nanoscale Research Letters. 2011;**6**(1):15. DOI: 10.1007/s11671-010-9749-0

[18] Pridgen EM, Alexis F, Farokhzad OC. Mint: Polymeric nanoparticle drug delivery technologies for oral delivery applications. Expert Opinion on Drug Delivery. 2015;**12**(9):1459–1473. DOI: 10.1517/17425247.2015.1018175

[19] Card JW, Magnuson BA. Mint: A review of the efficacy and safety of nanoparticle-based oral insulin delivery systems. American Journal of Physiology – Gastrointestinal and Liver Physiology. 2011;**301**:G956-G967

[20] Alai MS, Lin WJ, Pingale SS. Mint: Application of polymeric nanoparticles and micelles in insulin oral delivery. Journal of Food and Drug Analysis. 2015;**23**(3):351-358

[21] Gutjahr A, Phelip C, Coolen A-L, Monge C, Boisgard A-S, Paul S, Verrier B. Mint: Biodegradable polymeric Nanoparticles-Based vaccine adjuvants for lymph nodes targeting. Vaccines. 2016;**4**(4):34. DOI: 10.3390/vaccines4040034

[22] Ma W, Chen M, Kaushal S, McElroy M, Zhang Y, Ozkan C, Bouvet M, Kruse C, Grotjahn D, Ichim T, Minev B. Mint: PLGA nanoparticle-mediated delivery of tumorantigenic

peptides elicits effective immune responses. International Journal of Nanomedicine. 2012;7:1475-1487. DOI: 10.2147/IJN.S29506

[23] Farris E, Brown DM, Ramer-Tait AE, Pannier AK. Mint: Micro- and nanoparticulates for DNA vaccine delivery. Experimental Biology and Medicine (Maywood). 2016;**241**(9):919-929. DOI: 10.1177/1535370216643771

[24] Perez C, Sanchez A, Putnam D, Ting D, Langer R, Alonso MJ. Mint: Poly(lactic acid)-poly(ethylene glycol) nanoparticles as new carriers for the delivery of plasmid DNA. Journal of Controlled Release. 2001;**75**(1-2):211-224

[25] Bhavsar MD, Amiji MM. Mint: Gastrointestinal distribution and *in vivo* gene transfection studies with nanoparticles-in-microsphere oral system (NiMOS). Journal of Controlled Release. 2007;**119**(3):339-348

[26] Arun A, Reddy BSR. Mint: *In vitro* drug release studies from the polymeric hydrogels based on HEA and HPMA using 4-{(E)-[(3Z)-3-(4-(acryloyloxy)benzylidene)-2-hexylidene]methyl}phenyl acrylate as a crosslinker. Biomaterials. 2005;**26**:1185-1193

[27] Arun A, Reddy BSR. Mint: *In Vitro* drug release studies of 2-hydroxyethyl acrylate or 2-hydroxypropyl methacrylate_4-{(1E,4E)-5-[4-(acryloyloxy)phenyl]-3-oxopenta-1,4- dienyl}phenyl acrylate copolymer beads. Journal of Biomedical Materials Research Part B: Applied Biomaterials. 2005;**73**(2):291-300

[28] Phua K, Leong KW. Mint: Microscale oral delivery devices incorporating nanoparticles. Nanomedicine (London, England). 2010;**5**(2):161-163. DOI: 10.2217/nnm.09.113

[29] Ball RL, Knapp CM, Whitehead KA. Mint: Lipidoid nanoparticles for siRNA delivery to the intestinal epithelium: *In vitro* investigations in a Caco-2 model. PLoS One. 2015;**10**(7):e0133154. DOI: 10.1371/journal.pone.0133154

[30] Zhang Z, Tsai P-C, Ramezanli T, Michniak-Kohn BB. Mint: Polymeric nanoparticles-based topical delivery systems for the treatment of dermatological diseases. WIREs Nanomedicine and Nanobiotechnology. 2013;**5**:205-218. DOI: 10.1002/wnan.1211

[31] Reddy CU, Arun A, Amalraj A, Reddy BSR. Mint: Polymeric drugs based on sulphanilamide: Synthesis, antimicrobial studies and drug release studies. Journal of Pharmaceutical Sciences. 2007;**59**:1207-1213

[32] Bakrudeen HB, Sudarvizhi C, Reddy BSR. Mint: Starch nanocrystals based hydrogels: Construction, characterization and transdermal application. Materials Science and Engineering C: Materials for Biological Applications. 2016;**68**:880-889

[33] Nagarajan S, Reddy BSR, Tsibouklis J. Mint: *In vitro* effect on cancer cells: Synthesis and preparation of polyurethane membranes for controlled delivery of curcumin. Journal of Biomedical Materials Research Part A. 2011;**99A**:410-417

[34] Kilfoyle BE, Sheihet L, Zhang Z, Laohoo M, Kohn J, Michniak-Kohn BB. Mint: Development of paclitaxel-TyroSpheres for topical skin treatment. Journal of Controlled Release. 2012;**163**(1):18-24. DOI: 10.1016/j.jconrel.2012.06.021

[35] Ramezanli T, KilfoyleBE, Zhang Z, Michniak-Kohn BB. Mint: Polymeric nanospheres for topical delivery of vitamin D3. International Journal of Pharmaceutics. 2017;**516**(1-2): 196-203. DOI: 10.1016/j.ijpharm.2016.10.072

[36] Batheja P, Sheihet L, Kohn J, Singer AJ, Michniak-Kohn B. Mint: Topical drug delivery by a polymeric nanosphere gel: Formulation optimization and *in vitro* and *in vivo* skin distribution studies. Journal of Controlled Release. 2011;**149**(2):159-167. DOI: 10.1016/j.jconrel.2010.10.005

[37] Nagai N, Iwamae A, Tanimoto S, Yoshioka C, Ito Y. Mint: Pharmacokinetics and antiinflammatory effect of a novel gel system containing ketoprofen solid nanoparticles. Biological and Pharmaceutical Bulletin. 2015;**38**(12):1918-1924. DOI: 10.1248/bpb.b15-00567

[38] Cilurzo F, Selmin F, Gennari CG, Montanari L, Minghetti P. Mint: Application of methyl methacrylate copolymers to the development of transdermal or loco-regional drug delivery systems. Expert Opinion on Drug Delivery. 2014;**11**(7):1033-1045. DOI: 10.1517/17425247.2014.912630

[39] Rancan F, Amselgruber S, Hadam S, Munier S, Pavot V, Verrier B, Hackbarth S, Combadiere B, Blume-Peytavi U, Vogt A. Mint: Particle-based transcutaneous administration of HIV-1 p24 protein to human skin explants and targeting of epidermal antigen presenting cells. Journal of Controlled Release. 2014;**176**:115-122. DOI: 10.1016/j.jconrel.2013.12.022

[40] DeMuth PC, Min Y, Irvine DJ, Hammond PT. Mint: Implantable silk composite microneedles for programmable vaccine release kinetics and enhanced immunogenicity in transcutaneous immunization. Advanced Healthcare Materials. 2014;**3**(1):47-58. DOI: 10.1002/adhm.201300139

[41] Fonseca-Santos B, Gremião MPD, Chorilli M. Mint: Nanotechnology-based drug delivery systems for the treatment of Alzheimer's disease. International Journal of Nanomedicine. 2015;**10**:4981-5003. DOI: 10.2147/IJN.S87148

[42] Zhuang X, Xiang X, Grizzle W, Sun D, Zhang S, Axtell RC, Ju S, Mu J, Zhang L, Steinman L, Miller D, Zhang H-G. Mint: Treatment of brain inflammatory diseases by delivering exosome encapsulated Anti-inflammatory drugs from the nasal region to the brain. Molecular Therapy. 2011;**19**(10):1769-1779

[43] Zhuang X, Teng Y, Samykutty A, Mu J, Deng Z, Zhang L, Cao P, Rong Y, Yan J, Miller D, Zhang H-G. Mint: Grapefruit-derived nanovectors delivering therapeutic miR17 through an intranasal route inhibit brain tumor progression. Molecular Therapy. 2016;**24**(1):96-105

[44] Ong WY, Shalini SM, Costantino L. Mint: Nose-to-brain drug delivery by nanoparticlesin the treatment of neurological disorders. Current Medicinal Chemistry. 2014;**21**(37):4247-4256

[45] Sharma D, Maheshwari D, Philip G, Rana R, Bhatia S, Singh M, Gabrani R, Sharma SK, Ali J, Sharma RK, Dang S. Mint: Formulation and optimization of polymeric

nanoparticles for intranasal delivery of lorazepam using Box-Behnken design: *In vitro* and *in vivo* evaluation. BioMed Research International. 2014. Article ID 156010. DOI: 10.1155/2014/156010

[46] Chonkar A, Nayak U, Udupa N. Mint: Smart polymers in nasal drug delivery. Indian Journal of Pharmaceutical Sciences. 2015;**77**(4):367-375

[47] Park JS, Oh YK, Yoon H, Kim JM, Kim CK. Mint: *In situ* gelling and mucoadhesive polymer vehicles for controlled intranasal delivery of plasmid DNA. Journal of Biomedical Materials Research. 2002;**59**(1):144-151

[48] Nakamura K, Maitani Y, Lowman AM, Takayama K, Peppas NA, Nagai T. Mint: Uptake and release of budesonide from mucoadhesive, pH-sensitive copolymers and their application to nasal delivery. Journal of Controlled Release. 1999;**61**:329-335

[49] Aikawa K, Matsumoto K, Uda H, Tanaka S, Shimamura H, Aramaki Y. Mint: Hydrogel formation of the pH response polymer polyvinylacetaldiethylamino acetate (AEA). International Journal of Pharmaceutics. 1998;**167**:97-104

[50] Ensign LM, Cone R, Hanes J. Mint: Nanoparticle-based drug delivery to the vagina: A review. Journal of Controlled Release. 2014;**0**:500-514. DOI: 10.1016/j.jconrel.2014.04.033

[51] das Neves J, Araújo F, Andrade F, Michiels J, Ariën KK, Vanham G, Amiji M, Bahia MF, Sarmento B. Mint: *In vitro* and *Ex vivo* evaluation of polymeric nanoparticles for vaginal and rectal delivery of the Anti-HIV drug dapivirine. Molecular Pharmaceutics. 2013;**10**(7):2793-2807. DOI: 10.1021/mp4002365

[52] Woodrow KA, Cu Y, Booth CJ, Saucier-Sawyer JK, Wood MJ, Saltzman WM. Mint: Intravaginal gene silencing using biodegradable polymer nanoparticles densely loaded with small-interfering RNA. Nature Materials. 2009;**8**:526-533

[53] Steinbach JM, Weller CE, Booth CJ, Saltzman WM. Mint: Polymer nanoparticles encapsulating siRNA for treatment of HSV-2 genital infection. Journal of Controlled Release. 2012;**162**:102-110

[54] Li SF, Liu DX, Zheng YH, Yue Y, Huang YB, Jing XB. Mint: Inhibitory effects of paclitaxel-loaded PLA nanofibers against mice cervical cancers. Acta Metallurgica Sinica. 2012;**0**(9):1029-1034. DOI: 10.3724/SP.J.1105.2012.12066

[55] Zong S, Wang X, Yang Y, Wu W, Li H, Ma Y, Lin W, Sun T, Huang Y, Xie Z, Yue Y, Liu S, Jing X. Mint: The use of cisplatin-loaded mucoadhesive nanofibers for local chemotherapy of cervical cancers in mice. European Journal of Pharmaceutics and Biopharmaceutics. 2015;**93**:127-135. DOI: 10.1016/j.ejpb.2015.03.029

[56] Ordikhani F, Arslan ME, Marcelo R, Sahin I, Grigsby P, Schwarz JK, Azab AK. Mint: Drug delivery approaches for the treatment of cervical cancer. Pharmaceutics. 2016;**8**(3):23. DOI: 10.3390/pharmaceutics8030023

[57] Yoo JW, Giri N, Lee CH. Mint: pH-sensitive Eudragit nanoparticles for mucosal drug delivery. International Journal of Pharmaceutics. 2011;**403**:262-267

[58] Coco R, Plapied L, Pourcelle V, Jérôme C, Brayden DJ, Schneider YJ, Préat V. Mint: Drug delivery to inflamed colon by nanoparticles: Comparison of different strategies. International Journal of Pharmaceutics. 2013;**440**(1):3-12. DOI: 10.1016/j.ijpharm.2012.07.017

[59] Andonova V, Georgiev G, Toncheva V, Kassarova M. Mint: Preparation and study of Poly(vinyl acetate) and Poly(styrene) nanosized latex with Indometacin. Pharmazie. 2012;**67**(7):601-604

[60] Andonova VY, Georgiev GS, Georgieva VT, Petrova NL, Kassarova M. Mint: Indomethacin nanoparticles for applications in liquid ocular formulations. Folia Medica. 2013;**55**(1): 76-82

[61] Andonova V, Georgiev G, Toncheva V, Petrova N, Karashanova D, Penkov D, Kassarova M. Mint: Indomethacin loading and *in vitro* release properties from vinyl acetate homo- and co-polymer nanoparticles, coated with polyzwitterion and carbopol® shells. International Journal of Pharmaceutics. 2014;**6**(1):691-699

[62] Andonova V, Georgiev G, Toncheva V, Karashanova D, Katsarov P, Kassarova M. Mint: Carbopol® and chitosan coated nanoparticles with in-situ loaded indomethacin. American Journal of PharmTech Research. 2014;**4**(1):664-678

[63] Andonova V, Draganov M, Feodorova Y, Georgiev G, Kassarova M. Mint: Biological activity of Indomethacin-Loaded nanoparticles. World Journal of Pharmaceutical Research. 2014;**4**(1):85-101

[64] Andonova V, Zagorchev P, Katsarov P, Kassarova M. Mint: Eye drops with nanoparticles as drug delivery systems. International Journal of Pharmaceutical Sciences and Research. 2015;**7**(2):431-435

[65] Andonova V, Georgiev G, Dimitrova S, Katsarova M, Kassarova M. Mint: Characterization, *in vitro* evaluation and stability studies of Indomethacin-Loaded polyzwitterionic copolymer nanoparticles. International Journal of Drug Delivery Technology. 2015;**5**(3):89-97

[66] Almeida H, Amaral MH, Lobão P, Silva AC, Loboa JM. Mint: Applications of polymeric and lipid nanoparticles in ophthalmic pharmaceutical formulations: Present and future considerations. Journal of Pharmaceutical Sciences. 2014;**17**(3):278-293

[67] Giannavola C, Bucolo C, Maltese A, Paolino D, Vandelli MA, Puglisi G, Lee VH, Fresta M. Mint: Influence of preparation conditions on acyclovir loaded poly-d,l-lactic acid nanospheres and effect of PEG coating on ocular drug bioavailability. Pharmaceutical Research. 2003;**20**(4):584-590

[68] Kao HJ, Lin HR, Lo YL, Yu SP. Mint: Characterization of pilocarpine-loaded chitosan/carbopol nanoparticles. Journal of Pharmacy and Pharmacology. 2006;**58**(2):179-186

[69] Yuan XB, Yuan YB, Jiang W, Liu J, Tian EJ, Shun HM, Huang DH, Yuan XY, Li H, Sheng J. Mint: Preparation of rapamycin-loaded chitosan/PLA nanoparticles for immunosuppression in corneal transplantation. International Journal of Pharmaceutics. 2008;**349**(1-2):241-248

[70] Ibrahim HK, El-Leithy IS, Makky AA. Mint: Mucoadhesive nanoparticles as carrier systems for prolonged ocular delivery of gatifloxacin/prednisolone bitherapy. Molecular Pharmaceutics. 2010;**7**(2):576-585

[71] Gupta H, Aqil M, Khar RK, Ali A, Bhatnagar A, Mittal G. Mint: Sparfloxacin-loaded PLGA nanoparticles for sustained ocular drug delivery. Nanomedicine: Nanotechnology, Biology and Medicine. 2010;**6**(2):324-333

[72] Gupta H, Aqil M, Khar RK, Ali A, Bhatnagar A, Mittal G. Mint: Biodegradable levofloxacin nanoparticles for sustained ocular drug delivery. Journal of Drug Targeting. 2011;**19**(6):409-417

[73] Gupta H, Aqil M, Khar RK, Ali A, Bhatnagar A, Mittal G. Mint: Nanoparticles laden in situ gel for sustained ocular drug delivery. Journal of Pharmacy and Bioallied Sciences. 2013;**5**(2):162-165

[74] Gupta H, Aqil M, Khar RK, Ali A, Bhatnagar A, Mittal G. Mint: Nanoparticles laden in situ gel of levofloxacin for enhanced ocular retention. Drug Delivery. 2013;**20**(7):306-309

[75] Kumar D, Jain N, Gulati N, Nagaich U. Mint: Nanoparticles laden in situ gelling system for ocular drug targeting. Journal of Advanced Pharmaceutical Technology & Research. 2013;**4**(1):9-17

[76] Nasr FH, Khoee S, Dehghan MM, Chaleshtory SS, Shaffie S. Mint: Preparation end evaluation of contact lenses embedded with polycaprolactone-based nanoparticles for ocular drug delivery. Biomacromolecules. 2016;**17**(2):485-495

[77] Khurana G, Arora S, Pawar PK. Mint: Ocular insert for sustained delivery of gatifloxacinsesquihydrate: Preparation and evaluations. International Journal of Pharmaceutical Investigation. 2012;**2**(2):70-77. DOI: 10.4103/2230-973X.100040

[78] Thakur R, Swami G, Rahman M. Mint: Development and optimization of controlled release bioerodable anti infective ophthalmic insert. Current Drug Delivery. 2014;**11**(1):2-10

[79] O'Rourke M, Hanes J. Company profile: GrayBug, LLC. OnDrug Delivery 2014;**48**:8-9. Available from: www.ondrugdelivery.com/publications/48/Issue_48_Lo_Res.pdf [Accessed: 3 July 2017]

[80] Navratil T, Maynor B, Yerxa B. Improving outcomes in ophthalmology via sustained drug delivery. OnDrug Delivery. 2014;**48**:10-13. Available from: www.ondrugdelivery.com/publications/48/Issue_48_Lo_Res.pdf [Accessed: 3 July 2017]

[81] Rupenthal ID. Ocular drug delivery technologies: Exiting times ahead. OnDrug Delivery. 2015;**54**:7-11. Available from: http://www.ondrugdelivery.com/publications/54/Issue_54_Lo_Res.pdf [Accessed: 3 July 2017]

[82] Andonova V. Mint: A new direction in ophthalmic development: Nanoparticle drug delivery systems. Current Pharmaceutical Design. 2016;**22**(41):6313-6329

[83] Boddupalli BM, Mohammed ZN, Nath RA, Banji D. Mint: Mucoadhesive drug delivery system: An overview. Journal of Advanced Pharmaceutical Technology & Research. 2010;**1**:381-387

[84] Mašek J, Lubasová D, Lukáč R, Turánek-Knotigová P, Kulich P, Plocková J, Mašková E, Procházka L, Koudelka Š, Sasithorn N, Gombos J, Bartheldyová E, Hubatka F, Raška M, Miller AD, Turánek J. Mint: Multi-layered nanofibrous mucoadhesive films for buccal and sublingual administration of drug-delivery and vaccination nanoparticles – important step towards effective mucosal vaccines. Journal of Controlled Release. 2016;**249**:183-195

[85] Rana P, Murthy RSR. Mint: Formulation and evaluation of mucoadhesive buccal films impregnated with carvedilol nanosuspension: A potential approach for delivery of drugs having high first-pass metabolism. Drug Delivery. 2013;**20**(5):224-235

[86] Sapre AS, Parikh RK. Mint: Design of a buccal mucoadhesive, nanoparticles based delivery system of fluoxetine. Journal of Pharmaceutical Science and Bioscientific Research. 2012;**2**(3):148-161

[87] Al-Dhubiab BE. Mint: *In vitro* and *in vivo* evaluation of nano-based films for buccal delivery of zolpidem. Brazilian Oral Research. 2016;**30**(1):e126. DOI: 10.1590/1807-3107BOR-2016.vol30.0126

[88] Paranjpe M, Christel C. Müller-Goymann. Mint: Nanoparticle-Mediated pulmonary drug delivery: A review. International Journal of Molecular Sciences. 2014;**15**:5852-5873. DOI: 10.3390/ijms15045852

[89] El-Sherbiny IM, El-Baz NM, Yacoub MH. Mint: Inhaled nano- and microparticles for drug delivery. Global Cardiology Science & Practice. 2015;**2015**:2. DOI: 10.5339/gcsp.2015.2

[90] Doan TV, Couet W, Olivier JC. Mint: Formulation and *in vitro* characterization of inhalable rifampicin-loaded PLGA microspheres for sustained lung delivery. International Journal of Pharmaceutics. 2011;**414**(1-2):112-117. DOI: 10.1016/j.ijpharm.2011.05.007

[91] Reinbold J, Hierlemann T, Hinkel H, Müller I, Maier ME, Weindl T, Schlensak C, Wendel HP, Krajewski S. Mint: Development and *in vitro* characterization of poly(lactide-co-glycolide) microspheres loaded with an antibacterial natural drug for the treatment of long-term bacterial infections. Drug Design, Development and Therapy. 2016;**10**:2823-2832

[92] Chen L, Nakano K, Kimura S, Matoba T, Iwata E, Miyagawa M, Tsujimoto H, Nagaoka K, Kishimoto J, Sunagawa K, Egashira K. Mint: Nanoparticle-mediated delivery of pitavastatin into lungs ameliorates the development and induces regression of monocrotaline-induced pulmonary artery hypertension. Hypertension. 2011;**57**(2):343-350. DOI: 10.1161/HYPERTENSIONAHA.110.157032

[93] Anton N, Jakhmola A, Vandamme TF. Mint: Trojan microparticles for drug delivery. Pharmaceutics. 2012;**4**(1):1-25. DOI: 10.3390/pharmaceutics4010001

[94] Xu H, Hou Z, Zhang H, Kong H, Li X, Wang H, Xie W. Mint: An efficient Trojan delivery of tetrandrine by poly(N-vinylpyrrolidone)-block-poly(ε-caprolactone) (PVP-b-PCL) nanoparticles shows enhanced apoptotic induction of lung cancer cells and inhibition of its migration and invasion. International Journal of Nanomedicine. 2014;**9**:231-242. DOI: 10.2147/IJN.S55541

[95] Kaye RS, Purewal TS, Alpar HO. Mint: Simultaneously manufactured nano-in-micro (SIMANIM) particles for dry-powder modified-release delivery of antibodies. Journal of Pharmaceutical Sciences. 2009;**98**(11):4055-4068. DOI: 10.1002/jps.21673

[96] Bao W, Liu R, Wang Y, Wang F, Xia G, Zhang H, Li X, Yin H, Chen B. Mint: PLGA-PLL-PEG-Tf-based targeted nanoparticles drug delivery system enhance antitumor efficacy via intrinsic apoptosis pathway. International Journal of Nanomedicine. 2015;**10**:557-566. DOI: 10.2147/IJN.S75090

[97] Liu R, Wang Y, Li X, Bao W, Xia G, Chen W, Cheng J, Xu Y, Guo L, Chen B. Mint: Synthesis and characterization of tumor-targeted copolymer nanocarrier modified by transferrin. Drug Design, Development and Therapy. 2015;**9**:2705-2719. DOI: 10.2147/DDDT.S80948

[98] Xie J, Zhang X, Teng M, Yu B, Yang S, Lee RJ, Teng L. Mint: Synthesis, characterization, and evaluation of mPEG-SN38 and mPEG-PLA-SN38 micelles for cancer therapy. International Journal of Nanomedicine. 2016;**11**:1677-1686. DOI: 10.2147/IJN.S103110

[99] Kowalczuk A, Stoyanova E, Mitova V, Shestakova P, Momekov G, Momekova D, Koseva N. Mint: Star-shaped nano-conjugates of cisplatin with high drug payload. International Journal of Pharmaceutics. 2011;**404**(1-2):220-230

[100] Stoyanova E, Mitova V, Shestakova P, Kowalczuk A, Momekov G, Momekova D, Marcinkowski A, Koseva N. Mint: Reversibly PEGylated nanocarrier for cisplatin delivery. Journal of Inorganic Biochemistry. 2013;**120**:54-62

[101] Xia H, Gao X, Gu G, Liu Z, Hu Q, Tu Y, Song Q, Yao L, Pang Z, Jiang X, Chen J, Chen H. Mint: Penetratin-functionalized PEG-PLA nanoparticles for brain drug delivery. International Journal of Pharmaceutics. 2012;**436**(1-2):840-850. DOI: 10.1016/j.ijpharm.2012.07.029

[102] Shamma RN, Elkasabgy NA, Mahmoud AA, Gawdat SI, Kataia MM, Abdel Hamid MA. Mint: Design of novel injectable in-situ forming scaffolds for non-surgical treatment of periapical lesions: *In-vitro* and *in-vivo* evaluation. International Journal of Pharmaceutics. 2017;**521**(1-2):306-317

[103] Nagarajan S, Kiran MS, Tsibouklis J, Reddy BSR. Mint: Multifunctional star-shaped polylactic acid implant for potential use in angioplasty. Journal of Materials Chemistry B. 2014;**2**:6549-6559

Chapter 6

Spectroscopic Investigations on Polyvinyl Alcohol Film with Complex of Terbium Ions along with Bismuth Nanoparticles for Improved Green Emission

Gagandeep Kaur, Brijesh Kumar and S.B. Rai

Additional information is available at the end of the chapter

http://dx.doi.org/10.5772/intechopen.70274

Abstract

In this chapter, bismuth nanoparticles (NPs) have been synthesized by the pulsed laser ablation technique at different pH in different aqueous solutions (namely, water (H), water + sodium hydroxide (HN), and water + hydrochloric acid (HC)). The NPs in aqueous solutions have been characterized by transmission electron microscopy (TEM) and UV-Vis-NIR absorption techniques. The NPs are spherical, core shell, and hollow spheres in H, HN, and HC, respectively. The vibrational features have been studied using Raman technique and correlated with solid NPs, hollow NPs, core shell NPs, and NPs complex, etc. The Bi NPs were subsequently scattered with Tb^{3+} ions and their complex with salicylic acid (Sal) and 1,10-phenanthroline in aqueous solution of polyvinyl alcohol to get thin polymer films. Then photoluminescence properties of Tb^{3+} ions and the ($Tb(Sal)_3(Phen)$) complex were studied using 266 nm and 355 nm as excitation wavelengths to seek into the influence of Bi NPs on their emissive properties. Terbium ions in case of ($Tb(Sal)_3(Phen)$) complex together with NPs demonstrate an intense and extended emission spectrum in the 375–700 nm range for transition arising even from $^5D_3$ and $^5D_4$ levels to different ^7F_J levels on 266 nm excitation. Alternatively, the luminescence intensity of Tb^{3+} ions complexed with Sal and Phen in the thin polymer films is improved appreciably as compared with Tb^{3+} ions in the presence of Bi NPs on excitation with 355 nm.

Keywords: nanoparticles, laser processing, polymers, luminescence, thin films

1. Introduction

Optically active materials doped with lanthanide (RE) ions dispersed in polymer are appropriate for use in a diversity of optoelectronic applications such as amplifiers, fibers, and waveguides due to their stumpy expenditure and simplicity of processing and lofty performance [3, 13]. RE ions offer intense narrow spectral emissions from the long-lived excited-states arising from partially filled $4f^n$ orbitals [19]. When supplemented with the polymer host, these favorable radiative properties of RE ions have a tendency to get diminished by the intrinsic soaring vibrational frequencies of different bonds of polymers. Moreover, RE ion salts exhibit limited solubility in polymers and tend to aggregate, resulting in enhanced scattering and luminescence quenching even at low concentrations. Consequently, RE ions are united with an organic ligand before being dispersed in polymeric hosts to circumvent this. These organic ligands absorb the incident light energy and transfer it to the RE ion and can act as antenna ligands as shown in **Figure 1**.

The encapsulation of the RE ion is done with an appropriate organic ligand that is capable of transferring the input energy absorbed by the ligand to the RE ion. The triply ionized RE ions can be complexed with diverse types of strongly absorbing chelating ligands, e.g., β-diketones bipyridines, cryptands, calixarenes, cyclodextrins, and crown ethers [23]. Such complexed systems have large solubility in a lot of common organic solvents, such as chloroform, benzene, toluene, and are also soluble in several polymer matrices. One such ligand is salicylic acid (here after Sal), which is naturally prevalent as its glucosides in willow leaves and poplar, and birch. It is commonly recognized as ortho hydroxybenzoic acid or 2-hydroxybenzoic acid. Major component from the manufacture of salicylic acid is used in the production of aspirin (acetylsalicylic acid) nowadays. Molecular structure of salicylic acid is shown in **Figure 2**. The crystal structure of salicylic acid is monoclinic and has been resolved by Cochran. The crystal structure of salicylic acid is centro-symmetric carboxylic acid dimers. The hydroxyl group is hydrogen bonded intra-molecularly to the carbonyl oxygen. This leads to a less flexible molecule and dimer and a reduced intermolecular hydrogen bonding. The sensitized luminescence of lanthanide ions (especially Tb^{3+} ion by salicylic acid (Sal) has been found to play an important role in the analytical chemistry for the analysis of trace amount of Sal and its derivatives in biological systems. Kaur et al. [20] have prepared terbium complexes with salicylic acid and observed enhanced luminescence properties of terbium ion. RE(III) ions are chelated with ligands that have broad intense absorption bands. In these systems, intense ion luminescence originates from the intramolecular energy transfer

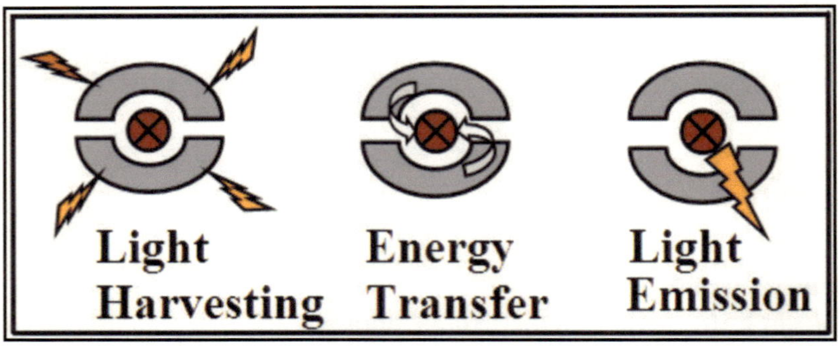

Figure 1. Pictorial representation of antenna effect.

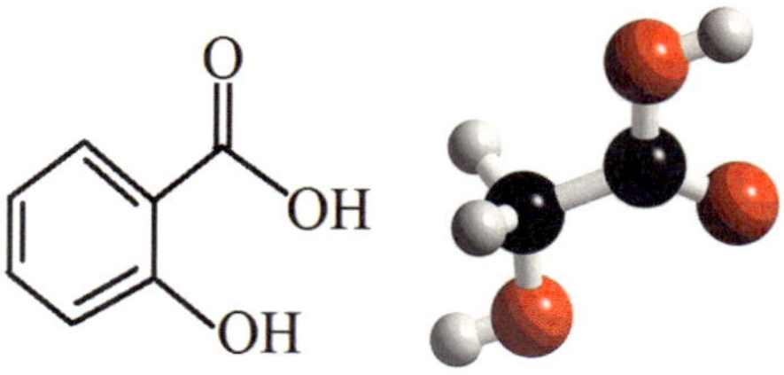

Figure 2. The structure of salicylic acid.

through the excited state of the ligand to the emitting level of the RE(III) ion. Salicylic acid is an effective aromatic carboxylic acid, also known as ligand transferring energy to the lanthanide ions. The sensitization of Tb(III) luminescence by the salicylic acid and its derivatives were also used for determination of salicylic acid in biological applications.

Thus, RE coordinated complexes in polymer hosts are used for numerous practical applications and their luminescence efficiency may be improved by a variety of methods [3, 4, 13, 20, 22]. Besides the sensitization by organic ligands, the fluorescence of the lanthanides in the complexes can be further enhanced by the use of synergistic agents, which provide an insulating layer around the lanthanide complex, reducing the probability of radiationless energy transfer from the complex to the solvent. In addition, these are usually adopted to expel adsorbed water from the first coordination sphere and thus enhance luminescence as water molecules quench the luminescence through radiationless deactivation. Recently, synergetic ligands have been used to control the supermolecular structure of rare earth complexes and grafted the RE complex in the host matrix to form homogeneous stable functional materials and different synergetic ligands such as trioctylphosphine oxide (TOPO), thenoyltrifluoroacetonatone (TTA), tripenylphosphine oxide (TPPO), 1,10-phenanthroline (Phen), 2,2-bipy (Dipy), trioctylphosphine (TOPO), etc. We have used 1,10-phenanthroline (Phen) for our investigations—10-phenanthroline (Phen) is a heterocyclic organic compound and a bidentate ligand. It forms a strong complex with most metal ions. The molecular formula of Phen is $C_{12}H_8N_2$, and the chemical structures of some such ligands are shown in **Figure 3**. Any effort to improve the efficiency by increasing the RE concentration does not succeed, as aggregation of ions takes place at higher concentrations and these acts as quenching centers.

Nanoparticles (NPs) are emerging as interesting luminescent nanoscale materials not only for basic research but also for numerous applications in varied devices such as optical amplifiers, color displays, solid-state lasers, etc. The NPs affect the luminescence and dynamics of optically energetic lanthanide ions. Occurrence of enhanced intensity of emission is correlated

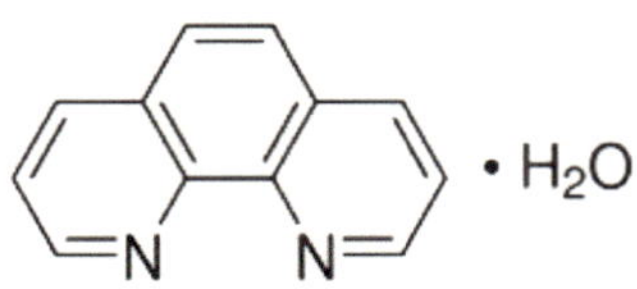

Figure 3. The structure of 1,10-phenanthroline ligand.

by means of energy transfer amid nanoparticles and the active rare earth ions. The second available option for the augmentation effect is via local energy field of NPs acting on the lanthanide ions positioned in their propinquity. This effect is prominent in particular when there is a matching of the surface plasmon resonance wavelength of the NPs and the incident light beam wavelength or luminescence wavelength [7, 26]. High surface-to-volume ratio and local phenomena such as absorption or changes in the surface electronic state may contribute significantly to special properties.

Metallic and semimetallic NPs are striking examples to be explored nowadays. The phenomenon of quantum size effect has a great influence on the physical properties that are very different from those of the bulk ones. Bismuth is a semimetal with a small energy overlap between the conduction and the valence bands. The size-induced semimetal to semiconductor transition and the related quantum confinement effects are potentially useful for optical and electro-optical device applications [2, 10, 16, 36, 37]. It has a high carrier mobility, a highly anisotropic Fermi surface, and miniature effective mass [2, 24, 46]. Below the size of 30 nm, the Bi NPs behave as semiconductors [44]. The formations of semimetal to semiconductor materials are being used in different scientific devices such as in optic and electro optic devices [39]. Bi NPs exhibit absorption in the UV region [27–29, 43]. The absorption peak shifts toward lower wavelengths in the case of smaller Bi NPs and, vice versa. The stability of Bi NPs decreases with a decrease in its particle size (due to enhanced surface to volume ratio). Nevertheless, these demonstrate strong reactive morphology. There are various methods to produce NPs. Among the diverse techniques used to fabricate NPs, laser ablation synthesis in solution (LASiS) generates pure NPs free from any type of contamination and thus, is matchless. In this technique, numerous factors, namely, laser power and wavelength, repetition rate of pulsed laser, spot size of laser beam, and most importantly the medium (solvent) used, etc., control the fabrication of NPs. The medium used for the preparation of NPs in LASiS decides the nature and steadiness of the colloidal NPs that can be further altered/improved by means of varying the pH of the solution. In numerous cases, laser-induced colloidal NPs are extremely reactive as they may react with the media and/or amidst themselves heading for the formation of multifarious configurations and varied agglomerated structures at room temperature provided there exists suitable liquid environment. But one can overcome this agglomeration by the inclusion of certain smart comparable ionic materials that could errand Coulomb repulsion [1], or else the NPs can be enclosed through dissimilar polarity layer that should favor Coulomb repulsion. The tuning of the structure of NPs from core shell to hollow nano structures is also achievable by pH variation. The researchers ought to use metallic nanoparticles for tuning the optical properties of the chosen host or activator (here the lanthanide ions), and for this intention, the sensitivity of NPs is crucial, which predominantly depends on the surface or volume plasmon frequency of NPs along with very negligible involvement of the local field effect. Thus, a deep knowledge and a clear understanding of the comprehensive mechanism of formation of NPs under diverse pH environments is a must. Bismuth-induced nanomaterials encompass exhaustive investigations to become the focal point of further research owing to their new applications (due to its semiconducting properties).

The inorganic luminescent rare earth ions and their stable complexes with varied compatible organic ligands enhance the emission of the RE ions in different media. Further, the incorporation of metallic or oxide NPs to the stable RE organic complexes unlocks the prospects for escalating their photoluminescence emission intensity by several orders of magnitude. Numerous researches and deep studies on the improvement of optical properties of RE ions and their complexes in diverse matrices have been approved and are still ongoing [34, 35, 38]. Polyvinyl alcohol (PVA) is a semicrystalline and nontoxic polymer possessing three isomers. Due to the different isomers, it shows a range of absorption and commensurate emission. The emissions are not significant from an application point of view and thus need to be enhanced. The incorporation of NPs comes to rescue and is the one best-suited option. This inclusion of NPs is achievable for the reason that the PVA molecules have large number of voids in their chains. Also, the occurrence of Bi NPs in the PVA molecule can transform/enhance the electronic emission from PVA by means of energy transfer from them and/or their local field effect. The augmentation is realistic for the reason that there exists the buffet of energy level of the PVA molecule as well as the Bi NPs absorbs immensely in the UV region. In the present chapter, differently shaped Bi NPs were prepared by laser ablation in solvents changing the pH of the solvents and characterized by transmission electron microscopy (TEM), SEM, and Raman studies. Laser ablation was used for preparation of Bi NPs in water, sodium hydroxide, and hydrochloric acid. The colloidal solution of as-prepared NPs were separately added to the Tb^{3+} ions and also to $(Tb(Sal)_3(Phen))$ complex in the PVA host polymer to attain thin films. The photoluminescence spectrum of the synthesized polymer films were scrutinized with 266 and 355 nm excitations. The reason for choosing these wavelengths being that the 266 nm excitation wavelength is resonant with the SPR band of Bi NPs, and the 355 radiation for excitation is off-resonant exciting only the Tb ions. Observed augmentation in the photoluminescence emission intensity of the activator terbium ion and its complex in the existence of bismuth NPs entrenched in the PVA host are explained.

2. Materials and experiments

All the ingredients terbium oxide, salicylic acid, and 1,10-phenanthroline used were all purchased from Sigma Aldrich. Tb_4O_7 and Sal were 99.9% pure. 1,10-phenanthroline was 99.5% pure. Bismuth plate with purity 99.0% was obtained from the same company and used for ablation and preparation of NPs.

2.1. Laser ablation technique to prepare Bi NPs

Laser ablation technique was used to prepare Bi NPs at different pH in different aqueous solutions (namely, water (H), water + sodium hydroxide (HN), and water + hydrochloric acid (HC)). **Figure 4** shows the experimental set up used for laser ablation for the production of Bi nanoparticles.

The irradiation source was the different wavelengths of an Nd:YAG laser with 7 ns pulse duration. The laser beam was tightly focused with the help of a short focal length (10 cm focal

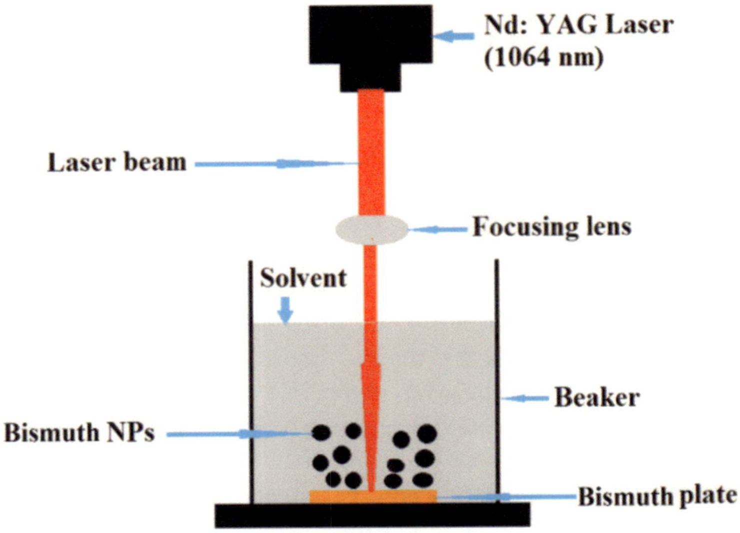

Figure 4. Experimental setup of laser ablation for preparation of Bi nanoparticles.

length) convergent lens on the copper plate (thickness: 0.6 mm, size: 18 × 30 mm^2) kept in a beaker with water 2.5 mm deep from the top surface. The liquid was continuously stirred using a magnetic stirrer. Laser ablation was continued for 45 min. The Bi plate was taken out of the liquid, and the sample was used for optical measurements and preparation of thin film samples. The same procedure was repeated for water + sodium hydroxide (HN), and water + hydrochloric acid.

2.2. Preparation of $Tb(Sal)_3(Phen)$ complex

0.010 mmol Tb_4O_7 was obtained by dissolving Tb_4O_7 in hydrochloric acid to obtain $TbCl_3$ solution. 0.015 mmol of salicylic acid and 0.005 mmol 1,10-phenanthroline were independently dissolved in 2.0 ml ethanol to get its ethanolic solution. Both were added and stirred rigorously for an hour to achieve $Tb(Sal)_3(Phen)$ complex as explained by Kaur et al. [21].

2.3. Preparation of polymer film with Tb^{3+} complex and Bi nanoparticles

Polyvinyl alcohol was dissolved in double distilled water to attain its 0.011 mmol transparent homogeneous solution. From the already-prepared laser ablated Bi nanoparticles in water, 5 ml of colloidal NPs were mixed separately with ethanolic solution of the as-prepared $Tb(Sal)_3(Phen)$ complex. The mixture was homogenized using a magnetic stirrer for 2 hours at room temperature and dispensed in aqueous solution of PVA. The resulting mixture was stirred thoroughly for 4–5 h and later poured in the polypropylene Petri dish and allowed to dry at its own without any heating agent to obtain the thin films. Flowchart depicting the steps for the preparation of polymer samples is shown in **Figure 5**.

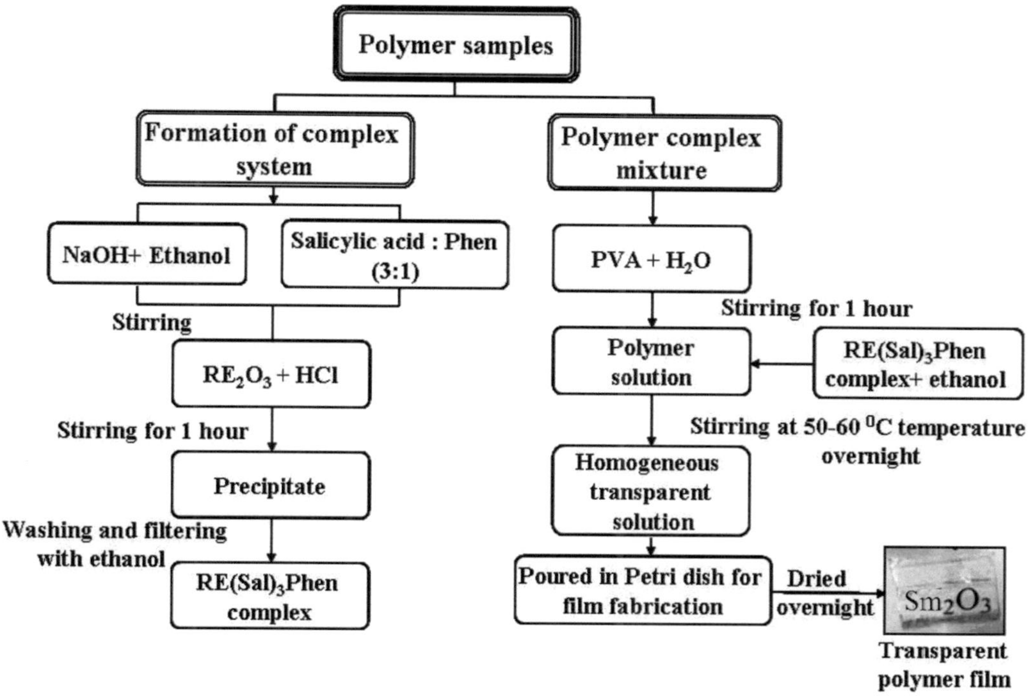

Figure 5. Flowchart depicting the steps for the preparation of polymer samples.

3. Results and discussions

3.1. Structural analysis using transmission electron microscopy (TEM)

TEM micrograph of Bi NPs in water, i.e., H (pH = 7) is shown in **Figure 6(a)**. The particles with average particle size of 14 nm and spherical in nature are observed. **Figure 6(b)** shows the size distribution of the particles as a histogram. The nanoparticles are poly-crystalline as seen from their selected area electron diffraction (SAED) patterns as shown in **Figure 6(c)**. Later, we supplemented it with significant amounts of acid and base to water separately to examine the consequence of pH on the size/dimension, contour/shape, and arrangement of NPs. At pH 9.7, namely, the medium is basic now, one monitors agglomerated core shell type NPs that are depicted in **Figure 6(d)**. An enlarged vision of core shell NPs is shown in its inset, verifying core diameter ~8 nm and shell thickness ~6 nm for the core shell NPs. **Figure 6(e)** demonstrates the histogram for its size distribution and the average size of NPs was found to be 18 nm. The SAED pattern for the same is given in **Figure 6(e)**. Here, the agglomeration of nanoparticles exists that leads to formation of bigger nanoparticles. This is due to the occurrence of the opposite polarity (Na^+ and OH^-) on the different NPs.

Spherical nanoparticles are formed when the pH of the sample is 7 (i.e., NPs prepared in pure water). Interestingly, when we add HCl in water and make the pH of the solution to 2.3,

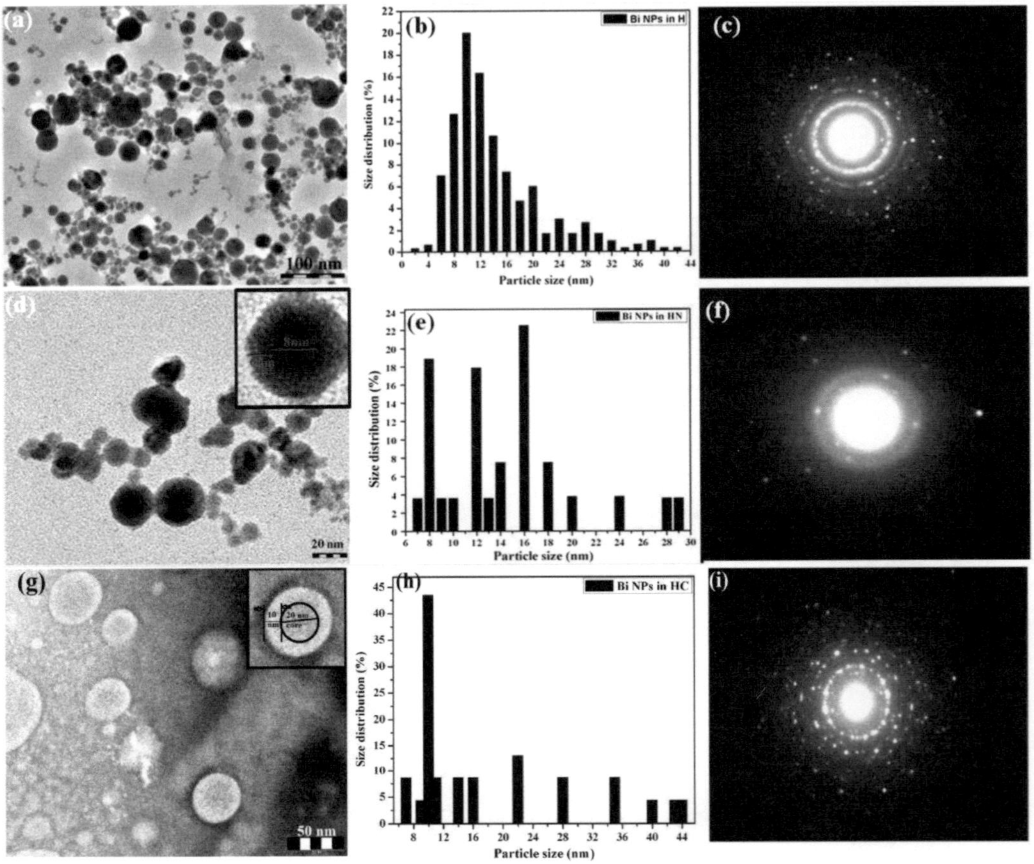

Figure 6. (a–c) TEM images of Bi NPs, particle size distribution, and SAED pattern in aqueous solution of H. (d–f) TEM images of Bi NPs, particle size distribution and SAED pattern in aqueous solution of HN. Inset of figure in (d) shows formation of core shell NPs. (g–i) TEM images of Bi NPs aqueous solution in water + HC with enlarged view of NPs with core size in the inset, particle size distribution and its SAED pattern (reproduced from Kumar et al. [48]).

the prepared NPs appear as hollow core shell NPs as shown in the TEM micrograph in **Figure 6(f)**. The average size of the NPs ranges between 15 and 45 nm (see **Figure 6(g)**). A few particles are bigger as they have swollen up. The diameter of the hollow core and thickness of shell is measured to be 20 and 10 nm, respectively, as shown in the inset of **Figure 6(g)**. **Figure 6(h)** shows the size distribution of the particles with average size of 21 nm. The SAED pattern presenting polycrystalline character is revealed in **Figure 6(i)**. All annotations provide evident confirmation that the shape of NPs depends on the pH of the used medium. The character of NPs transforms to hollow from core shell if we alter the pH and change it to acidic from the basic one.

The formation of hollow NPs can be understood on the basis of the Kirkendall effect [25]. The creation of bismuth nano particles initiate as soon as the laser beam is incident on the metallic plate of bismuth and focused carefully. Now the pH of the medium plays a crucial role. Spherical Bi NPs are formed in neutral pH. But the case differs for the acidic medium, as a few of the formed bismuth NPs react with the medium to produce bismuth oxychloride (BiOCl) that gets coated on the bismuth core NPs with the passage of time of ablation. It is noteworthy that the diffusion of bismuth ions is exceedingly rapid when compared to that of chloride ions. The difference in the diffusion rate generates vacancies in the core of bismuth NPs, and when these vacancies are formed in surplus, these join together to form voids. Because of these analogous reasons and facts, core-shell configuration is commonly observed to form voids in larger NPs. Even the created voids are not proportional in every case. This is probably owing to formation of partial hollow nanospheres as observed in CdS/Cd hollow NPs by researchers [5].

One does not observe the formation of hollow NPs when pH of media changes to basic. The reason accounted for this is possibly that in this case, bismuth hydroxide forms the shell and gets deposited on the bismuth core. Here in this case, the diffusivity of hydroxide ion is greatly better than that of bismuth ion. Consequently, there is a superior possibility of formation of vacancies in the core to acquire a core-shell type structure. This, moreover, augments the density of particles in the core and diminishes their amount in the shell.

3.2. Structural study using scanning electron microscopy (SEM)

Bi NPs in aqueous solution of H, HN, and HC, respectively, were added to the polymer films and their SEM images are shown in **Figure 7(a–c)**. The nanoparticles appear embedded in the case of water and basic medium, but in acidic media, some of the bismuth NPs reacts with acid to form flower-like clusters of BiOCl that are quite obvious in **Figure 7(c)**.

3.3. Raman measurement

To understand the formation of Bi NPs and their compounds attributed to different environments (H, HN, and HC), the Raman spectra of colloidal Bi NPs were measured. Raman spectrum of Bi NPs in aqueous solution of water (H), water + NaOH (HN), and water + HCl (HC),

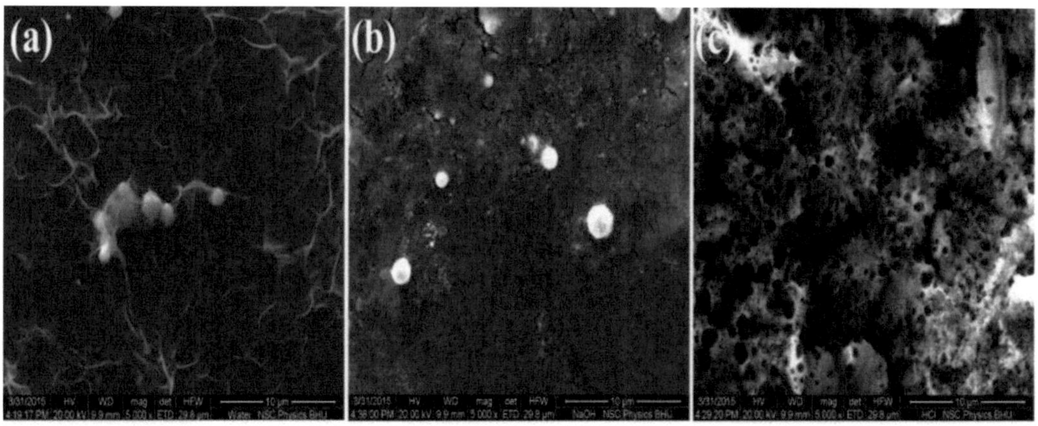

Figure 7. SEM images of Bi NPs in polymer films with H (a), HN (b) and HC (c) respectively (reproduced from Kumar et al. [48]).

respectively, in the range 50–400 cm^{-1} are depicted in **Figure 8(a)–(c)**. The spectra demonstrate numerous active vibrational modes in this frequency range. These bands are customized by the environment and bestow lucid information about molecules formed in different media.

Figure 8(a) shows the Raman vibrational peaks owing to Bi NPs in water. Two vibrational bands at 66.48 and 92.51 cm^{-1} are observed that can be attributed to pure Bi NPs in the host [11, 18, 30, 31]. Both these peaks depend upon the particle size. The former lower frequency band is assigned to the E_g mode and the later corresponds to A_{1g} mode. Generally, the vibrational frequencies of A_{1g} mode of Bi NPs lie between 85 and 95 cm^{-1}, and the Raman frequencies of the lower E_g mode lie between 59 and 75 cm^{-1}, which get somewhat modified with the size of the nanoparticles and transfer toward lower frequencies for smaller nanoparticles.

But the case is very different on changing the pH of the medium by adding NaOH (HN). **Figure 8(b)** depicts the Raman vibrational peaks of Bi NPs in HN. The peaks for the E_g and A_{1g} modes in the basic medium shift to 65.25 and 92.73 cm^{-1} with full width at half maxima 9.86 and 10.51 cm^{-1}, respectively. Two other peaks also emerge at 121.82 and 312.56 cm^{-1} in the spectrum. These Raman bands are assigned to Bi—O stretching vibrations confirming that the species formed as nanoparticles are due to α-Bi_2O_3 [8, 15]. Similarly, by decreasing the pH of the medium on the addition of considerable quantity of HCl to make it acidic, namely HC, the intensity of the Raman bands gets reduced when compared with the bands of Bi NPs in H and HN (see **Figure 8(c)**). In this case, an intense peak is seen at 141.24 cm^{-1} along with weak

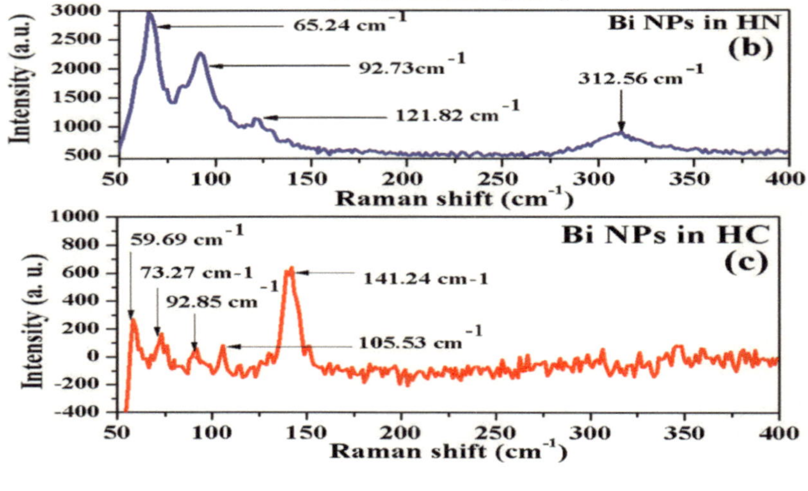

Figure 8. (a–c) Raman spectrum of Bi NPs in aqueous solution of water (H), water + NaOH (HN), and water + HCl (HC), respectively (reproduced from Kumar et al. [48]).

bands observed at 59.69, 73.27, 92.85, and 105.53 cm^{-1}. The weaker Raman bands are assigned to the E_g and A_{1g} modes of vibration for Bi in α-Bi_2O_3. The band observed at 105.53 cm^{-1} may be attributed to Bi for α-Bi_2O_3 phase [12, 32]. The two peaks observed in HN at 121.82 and 312.56 cm^{-1} disappear here. This shows that the vibrational modes of Bi in α-Bi_2O_3 dominate over Bi—O vibration in α-Bi_2O_3. In this case (i.e., HC), two new Raman peaks are observed at 141.24 (stronger) and 59.69 cm^{-1} (weaker) due to BiOCl molecule [45, 47]. The former peak is attributed to internal stretching of Bi—Cl, and the later weaker one is assigned to its external stretching [39–41]. A weak band appears at 396 cm^{-1}, which is a consequence of the motion of the oxygen atom and designated as B_{1g} band, but it is not optically considerable in this case. It is necessary to mention that in the laser ablation synthesis in solution, momentous chemical processes occur after the ablation of the target. Consequently, the intensity of the peaks is correlated to the concentration of a particular species developed at that instant. Thus, one may infer that the intense band at 141.24 cm^{-1} due to BiOCl evidently suggests its larger concentration. The positions of the Raman peaks for Bi NPs in H, HN, and HC along with their full width at half maxima (FWHM) and their respective intensity are tabulated in **Table 1**.

3.4. UV-Vis absorption

Figure 9 shows the absorption spectra of Bi NPs in H, HN, and HC solutions. The absorption peak for Bi NPs in water emerges at 265 nm. This is accredited to the plasmonic peak frequency of Bi NPs. In the case of HN, two peaks are observed. The first peak at 233 nm is attributed to the plasmon frequency of Bi_2O_3 and the second peak at 274 nm is associated with that of Bi NPs. There is a shift in the Bi plasmon peak toward higher wavelength in this case as compared to the Bi plasmon peak in pure water (265 nm) that supports the agglomeration of Bi NPs. Absorption

Bi NPs in H	Raman peak position (cm^{-1})	66.48	92.51	—	—	—
	FWHM (cm^{-1})	7.49	5.78	—	—	—
	Normalized intensity	0.99	0.44	—	—	—
Bi NPs in HN	Raman peak position (cm^{-1})	65.25	92.73	121.82	312.56	—
	FWHM (cm^{-1})	9.86	10.51	3.71	33.06	—
	Normalized intensity	0.98	0.71	0.26	0.17	—
Bi NPs in HC	Raman peak position (cm^{-1})	59.69	73.26	92.85	105.53	141.24
	FWHW (cm^{-1})	2.86	4.89	4.04	2.71	9.01
	Normalized intensity	0.80	0.74	0.69	0.70	0.98

Table 1. Raman peaks position, FWHM, and normalized intensity of Bi NPs in water (H), water + NaOH (HN), and water + HCL (HC).

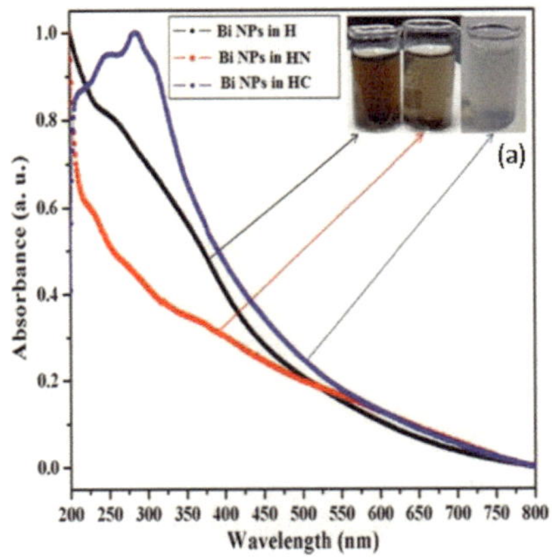

Figure 9. UV-Vis-NIR absorption spectrum of Bi NPs in aqueous solution of H, HN, and HC and inset figure shows their corresponding color (a) (reproduced from Kumar et al. [48]).

spectra for Bi NPs in HC sample show a weird behavior with absorption peaks at 245, 285, and 309 nm that are attributed to Bi_2O_3, Bi NPs, and BiOCl. The main plasmon peak is shifted slightly further toward higher wavelengths. The shifting of the absorption peak to the higher wavelength region is an indication of increased agglomeration of NPs. Wang et al. [41] have reported the absorption peak at 267 nm due to semimetal bismuth NPs in PVP solution. Creighton and Desmond [6] have reported the absorption band around 270–280 nm for bismuth particles of size ranging in 10 nm. This inconsistency in intensity of the absorption peaks appears as an outcome of the concentration of the species in solution. The inset to **Figure 9** shows the color of the colloidal solution after ablation. Different colors signify formation of Bi NPs and their conversion to additional forms, namely Bi_2O_3, BiOCl, or Bi NPs and its core shell structure.

The surface Plasmon resonance is the characteristic of NPs embedded in a dielectric host and is ascribed to combined oscillations of the electrons responding to the optical excitation energy. Optical absorption spectrum of Bi NPs prepared in water by laser ablation for 20 min is shown in **Figure 10**, which depicts its characteristic surface plasmon resonance peak at 267 nm together with a diminutive band at 283 nm that may be assigned to $^4S_{3/2} \rightarrow {}^2P_{3/2}$ of Bi^0 transitions [33]. On ablation of the bismuth target for 40 min, the peak at 267 nm is observed to diminish, and the intensity of the peak at 283 nm started increasing, which indicates that the Bi^{3+} ions are reduced completely to bismuth NPs [9, 17, 40, 41] Also, it is observed that the NPs get agglomerated giving bigger-sized NPs on ablation for a longer period and, hence, absorption at a longer wavelength. Gutierrez and Henglein [14] reported that nanometer-sized bismuth particles exhibited an absorption at ~253 nm, and according to Creighton and Desmond [6], the first absorption band of 10 nm bismuth particles should appear around 270–280 nm. Polyvinylpyrroldone-stabilized bismuth NPs have been synthesized by Wang et al. [41] with an absorption peak at 281 nm. Our result fits well with the above two literature values.

The absorption spectrum of ($Tb(Sal)_3(Phen)$) complex in PVA with and without laser-ablated Bi NPs are also shown in **Figure 10**. The ($Tb(Sal)_3(Phen)$) complex in the PVA film shows a band centered at 315 nm that may be attributed to the $S_0 \rightarrow S_1$ singlet state absorption of salicylic acid. Also, the absorption band due to $\pi \rightarrow \pi^*$ transition of PVA exists in this region, so

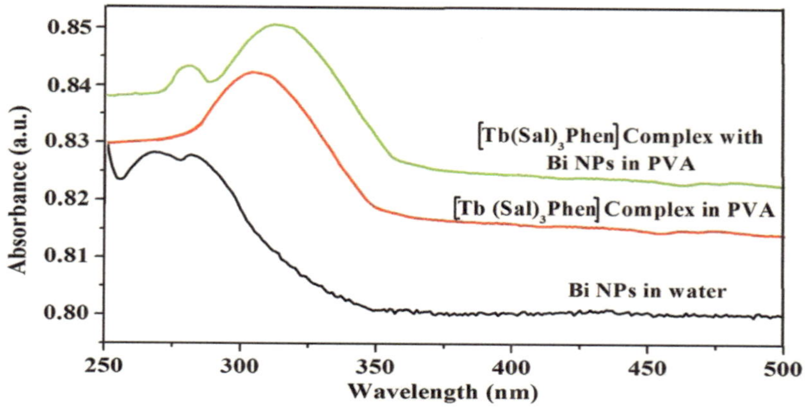

Figure 10. Absorption spectra of Bi NPs in water ablated for 20 min, (Tb(Sal)$_3$(Phen)) complex in PVA and (Tb(Sal)$_3$(Phen)) complex with Bi NPs in PVA (reproduced from Kaur et al. [24]).

there may be an overlapping of the absorption bands of PVA and Sal. The (Tb(Sal)$_3$(Phen)) complex with Bi NPs in PVA illustrates an absorption peak for Bi NPs along with the absorption band of Sal. It is observed that this band of Sal shows a red shift of ~8 nm on addition of Bi NPs along with the (Tb(Sal)$_3$(Phen)) complex in PVA film. The shift can be ascribed to aggregation of the complex through NPs.

4. Excitation spectra

The photoluminescence excitation spectra of Tb^{3+} ions in PVA, Tb^{3+} ions with Bi NPs in PVA and (Tb(Sal)$_3$(Phen)) complex with Bi NPs in PVA corresponding to the $^5D_4 \rightarrow ^7F_5$ transition of Tb^{3+} ion monitored at 544 nm were recorded and are shown in **Figure 11**.

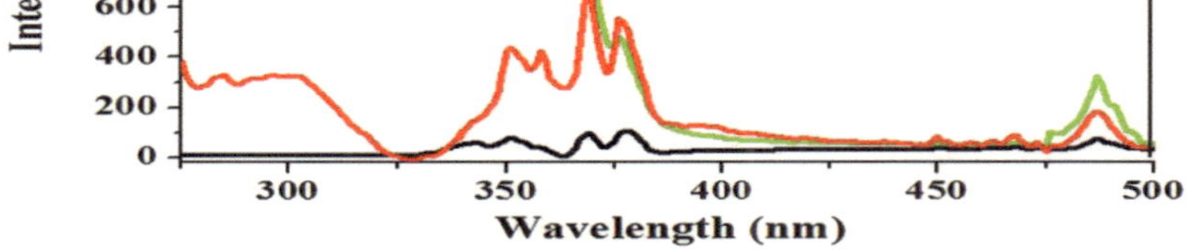

Figure 11. Excitation spectra of Tb^{3+} ions in PVA, Tb^{3+} ions with Bi NPs in PVA and the (Tb(Sal)$_3$(Phen)) complex with Bi NPs in PVA corresponding to the $^5D_4 \rightarrow ^7F_5$ transition of Tb^{3+} ions monitored at 544 nm (reproduced from Kaur et al. [24]).

The excitation spectrum of Tb^{3+} ion in PVA sample also shows a few weak bands at 341, 352, 358, 369, 377, and 488 nm wavelengths corresponding to absorption of Tb^{3+} ions. When Bi NPs were added to $TbCl_3$, the excitation intensity of the bands got improved. There also appears a weak broad plasmonic band for Bi NPs at 285 nm. The excitation spectrum $(Tb(Sal)_3(Phen))$ complex and laser-ablated Bi NPs in PVA depicts an extensive excitation band between 275 and 375 nm that may be attributed to $n \rightarrow \pi^*$ transition of the salicylate ligands. There seems to be appreciable enhancement in the intensity of the bands corresponding to the Tb^{3+} ion emission. This vividly signifies effectual sensitization of Tb^{3+} ions by the ligands pointing to a competent antenna effect [20].

5. Photoluminescence using 266 nm wavelength

The photoluminescence spectra depicts the emission bands of Tb^{3+} ions, Tb^{3+} ions with Bi NPs and the $(Tb(Sal)_3(Phen))$ complex with Bi NPs in PVA in the range of 375–700 nm on excitation with the SPR band of NPs using 266 nm radiation and is shown in **Figure 12**.

The spectra of Tb^{3+} ions exhibit characteristic emission peaks at 487, 544, 583, and 618 nm for Tb^{3+} ions emanating from $^5D_4 \rightarrow ^7F_J$ (J = 6, 5, 4, 3) transitions, respectively, and among them, the $^5D_4 \rightarrow ^7F_5$ transition (544 nm) is the most intense one. The emission intensity of Tb^{3+} bands was enhanced on incorporating Bi NPs, but the effect is more prominent in the case of the $(Tb(Sal)_3(Phen))$ complex in PVA as in this case, the emission emanating from the $^5D_3$ level also appear, which is an additional interesting feature.

The mechanism for augmentation of the emission intensity of the observed transitions may be elucidated with the help of a partial energy-level diagram showing different routes of excitation of Tb^{3+} ions, and the respective emissions are shown in **Figure 13**. Primarily, the 266 nm photon excites the $^5H_7$ level through the $^7F_6 \rightarrow ^5H_7$ absorption transition of Tb^{3+} ions. Then the excited Tb^{3+} ions relax nonradiatively down to $^5D_3$ and $^5D_4$ levels to yield the emissions from these level to lower lying levels (^7F_J; J = 1–6). This excitation radiation, i.e., 266 nm, moreover,

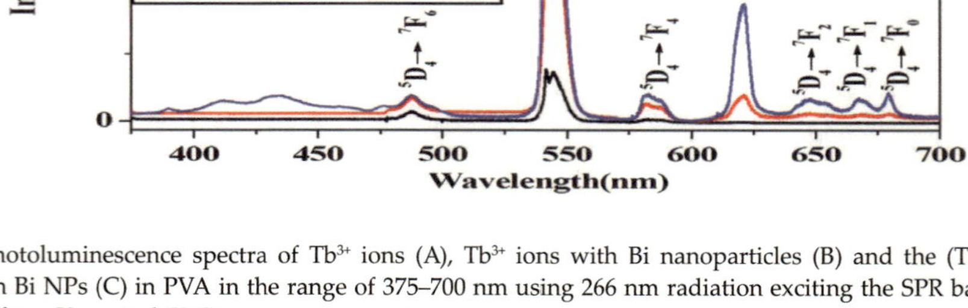

Figure 12. Photoluminescence spectra of Tb^{3+} ions (A), Tb^{3+} ions with Bi nanoparticles (B) and the $(Tb(Sal)_3(Phen))$ complex with Bi NPs (C) in PVA in the range of 375–700 nm using 266 nm radiation exciting the SPR band of Bi NPs (reproduced from Kaur et al. [24]).

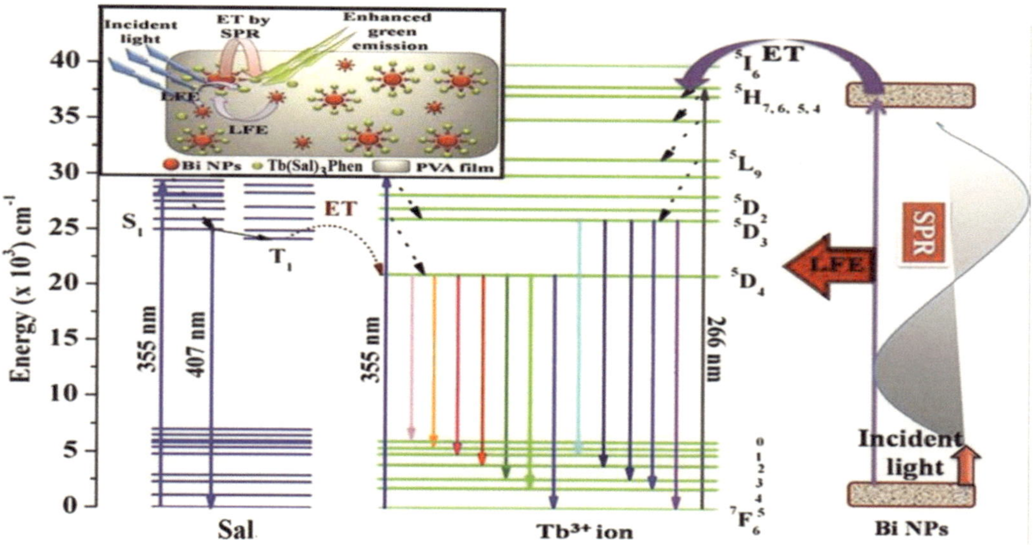

Figure 13. Partial energy-level diagram showing different routes of excitation of Tb^{3+} ions and the respective emissions exciting the SPR band of Bi NPs and the Sal ligand, respectively, on excitation with 266 and 355 nm radiations for improved green emission. The inset to the figure depicts Tb^{3+} ions, Bi NPs, and (Tb(Sal)$_3$(Phen)) complex in PVA host presenting the plasmonic and the field effect to boost the emission of Tb^{3+} ion. SPR refers to surface plasmon resonance, LFE stands for local field effect, and ET represents energy transfer (reproduced from Kaur et al. [24]).

excites the surface plasmon band of Bi NPs. This engrossed excitation energy from the surface plasmon band of Bi NPs is transferred to Tb^{3+} ions. It further improves the build up of population of higher $^5D_3$ and $^5D_4$ levels of Tb^{3+} ion. This is cause for enhancement in the photoluminescence emission intensity. The great absorption cross-section of the bismuth plasmon band grounds for an amplified excitation of $^5D_4$ level of Tb^{3+} ion by means of energy transfer from the excited surface plasmon resonance band of Bi NPs. It is noteworthy to mention that the ligand Sal does not absorb at 266 nm radiation as no energy level of Sal exists at this energy. It merely encapsulates the Tb^{3+} ion to cut it off from the host vibrations and intensifies the emission from the complexed Tb^{3+} ion, thus resulting in the emergence of emission from the $^5D_3$ level.

6. Photoluminescence using 355 nm wavelength

The photoluminescence spectra of Tb^{3+} ion, Tb^{3+} ion with Bi NPs and (Tb(Sal)$_3$(Phen)) complex with Bi NPs in PVA in the range between 375 and 700 nm using 355 nm excitation radiation, namely, nonresonant excitation is shown in **Figure 14**. The photoluminescence emission spectra is similar to the previous one, but the photoluminescence emission intensity for (Tb(Sal)$_3$(Phen)) complex with Bi NPs is enhanced to a large extent in the present case. This improvement in the photoluminescence emission intensity can be understood by the following mechanism.

This nonresonant 355 nm excitation radiation excites equally the Tb^{3+} ion in addition to the Sal ligand to their excited states. It should be mentioned here that the Bi NPs do not absorb this wavelength. This incident excitation energy is directly absorbed by the $^5L_9$ level of Tb^{3+} ion. It

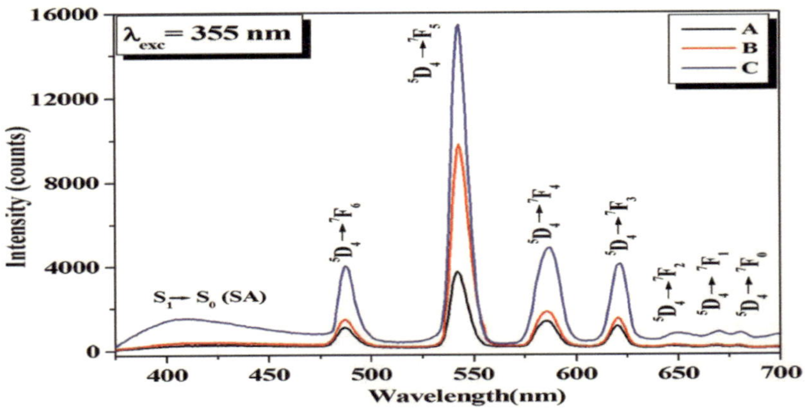

Figure 14. Photoluminescence spectra of Tb^{3+} ions (A), Tb^{3+} ions with Bi NPs (B), and $(Tb(Sal)_3(Phen))$ complex with Bi NPs (C) in PVA in the range of 375–700 nm on excitation with 355 nm radiation (reproduced from Kaur et al. [24]).

then relaxes nonradiatively and populates the emitting $^5D_4$ level. Along with this, the optical energy absorbed by the Sal ligand is also transferred to the resonating Tb^{3+} ions populating the $^5D_4$ level via intersystem crossing and the consequent energy transfer process that is the reason for enhancing the emission intensity. Also, the Bi NPs form a local plasmonic field around the $(Tb(Sal)_3(Phen))$ complex, and the high-field gradients of NPs increase the lifetime of the emitting $^5D_4$ level of Tb^{3+} ions [42]. The coupling between the radiative transitions, and the field effect is the fundamental basis for the enhancement in intensity as shown in the inset of **Figure 14**. The increase in the lifetime of the $^5D_4$ level of Tb^{3+} ion is clearly observed in the decay curves for the $^5D_4 \rightarrow {}^7F_5$ transition of Tb^{3+} ions in the presence and absence of Bi NPs (as seen in **Figure 15**). Herein, the point to mention is that different transitions of Tb^{3+} ion respond differently to the

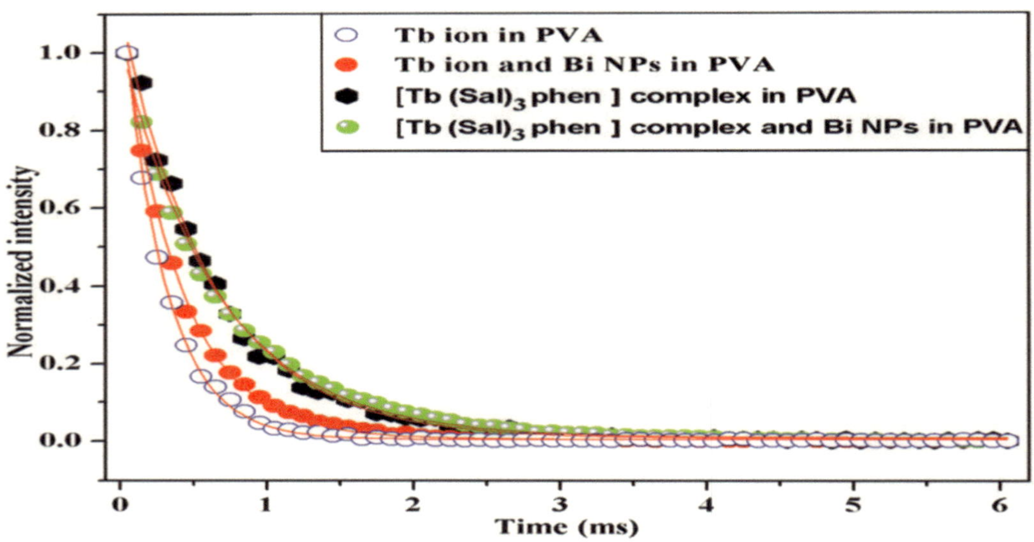

Figure 15. The decay curves for the $^5D_4 \rightarrow {}^7F_5$ transition at 544 nm of Tb^{3+} ions with and without NPs and the $(Tb(Sal)_3(Phen))$ complex with and without Bi NPs in PVA using 355 nm radiation (reproduced from Kaur et al. [24].

http://dx.doi.org/10.5772/intechopen.70274

local field gradients of Bi NPs. The reason behind this may be that magnetically allowed dipole transitions differ in interaction with the surface plasmon field of NPs than the electrically allowed dipole transitions. Nevertheless, the photoluminescence emission intensity of Tb^{3+} ion on excitation with 355 nm nonresonant radiation is larger to a great extent than that of 266 nm resonant excitation. It undoubtedly reveals that the transfer of energy to Tb^{3+} ions through the salicylic acid ligand is more proficient as compared with other channels of energy transfer.

7. Time-resolved photoluminescence spectroscopy

Figure 15 represents the decay curves for the $^5D_4 \rightarrow {}^7F_5$ transition monitored at 544 nm of Tb^{3+} ions with and without Bi NPs and the $(Tb(Sal)_3(Phen))$ complex with Bi NPs in PVA using 355 nm radiation.

The decay curves were fitted with first-order exponential fits that lead to a larger lifetime of $^5D_4$ level of Tb^{3+} ions in the presence of NPs. The values obtained for lifetimes are tabulated in **Table 2**.

This increase in lifetime is explained due to the local surface fields of Bi NPs as given in the inset of **Figure 13**. Thus, photoluminescence properties of polymer-doped RE complexes can be improved by adding NPs [22]. There is a substantial improvement in the lifetime values when Bi NPs are added to the $TbCl_3$ in PVA and, alternatively, for the $(Tb(Sal)_3(Phen))$ complex, there is just a small increase in the lifetime.

Sample	Lifetime (μs)
Tb^{3+} ions in PVA	287
Tb^{3+} ions with Bi NPs in PVA	399
$(Tb(Sal)_3(Phen))$ in PVA	625
$(Tb(Sal)_3(Phen))$ with Bi NPs in PVA	659

Table 2. The values for the lifetime of Tb^{3+} ions, Tb^{3+} ions with Bi NPs, $(Tb(Sal)_3(Phen))$ complex with Bi NPs, and $(Tb(Sal)_3(Phen))$ with Bi NPs in PVA.

8. Conclusion

To recapitulate, bismuth nanoparticles have been prepared by laser ablation technique at in different aqueous solutions, namely water, water + sodium hydroxide, and water + hydrochloric acid. TEM micrographs confirm the formation of spherical, core shell, and hollow spheres in H, HN, and HC, respectively, with variation in size. Further, the Bi NPs were subsequently scattered with Tb^{3+} ions and their complex with salicylic acid (Sal) and 1,10-phenanthroline in aqueous solution of polyvinyl alcohol to get thin polymer films. Then the photoluminescence properties of Tb^{3+} ions and the $(Tb(Sal)_3(Phen))$ complex were studied using 266 and 355 nm as excitation wavelengths. The emission efficiency of Tb^{3+} ions and their complex is seen to be enhanced in the presence of Bi NPs on excitation with both the radiations. On 266 nm excitation, a comprehensive photoluminescence emission spectrum of Tb^{3+} ions is observed for $Tb(Sal)_3(Phen)$ complex with Bi NPs spanning the region between 375 and 700 nm depicting transitions from $^5D_3/^5D_4$ levels to diverse ^7F_J levels. It is worthy to mention that the luminescence enhancement is better for ($Tb(Sal)_3(Phen)$) complex with 355 nm excitation radiation. The augmentation in intensity is ascribed to the coupling of plasmonic and local field effect of Bi nanoparticles on Tb^{3+} ion that influence the lifetime of radiative level of Tb^{3+} ion in addition to transfer of energy from Sal to Tb^{3+} ion.

Acknowledgements

G. Kaur acknowledges DST for the project grant in the form of SR/WOS-A/PS-57/2013 and RSC advances and Dalton Transactions for the reproduction of figures.

Author details

Gagandeep Kaur*, Brijesh Kumar and S.B. Rai

*Address all correspondence to: gagandeep_bhu@yahoo.com

Laser and Spectroscopy Laboratory, Department of Physics, Banaras Hindu University, Varanasi, India

References

[1] Amendola V, Meneghetti M. Laser ablation synthesis in solution and size manipulation of noble metal nanoparticles. Physical Chemistry Chemical Physics. 2009;**11**:3805-3821

[2] Balan L, Schneider R, Billaud D, Fort Y, Ghanbaja J. A new synthesis of ultrafine nanometre-sized bismuth particles. Nanotechnology. 2004;**15**:940-944

[3] Binnemans K. Lanthanide-based luminescent hybrid materials. Chemical Reviews. 2009; **109**;4283-4374

[4] Bekiari V, Lianos P. Multicolor emission from terpyridine lanthanide ion complexes encapsulated in nanocomposite silica/poly (ethyleneglycol) sol–gel matrices. Journal of Luminescence. 2003;**101**:135-140

[5] Cabot A, Smith RK, Yin Y, Zheng H, Reinhard BM, Liu H, Alivisatos AP. Sulfidation of cadmium at the nanoscale. ACS Nano. 2008;**2**:1452-1458

[6] Creighton JA, Desmond GE. Ultraviolet-visible absorption spectra of the colloidal metallic elements. Journal of the Chemical Society, Faraday Transactions. 1991;**87**(24):3881-3891

[7] Da Silva DM, Kassab LRP, Luthi SR, Araujo CB, Gomes ASL, Bell MJV. Applied Physics Letters. 2007;**90**:081913

[8] Denisov VN, Ivlev AN, Lipin AS, Mavrin BN, Orlov VG. Raman spectra and lattice dynamics of single-crystal α-Bi_2O_3. Journal of Physics Condensed Matter. 1997;**9**:4967-4978

[9] De Sande JCG, Missana T, Afonso CN. Optical properties of pulsed laser deposited bismuth films. Journal of Applied Physics. 1996;**80**:7023-7027

[10] Fang J, Stokes KL, Wiemann JA, Zhou WL, Dai J, Chen F, Connor CJO. Microemulsion-processed bismuth nanoparticles. Materials Science and Engineering B. 2001;**83**:254-257

[11] Gondal MA, Saleh TA, Drmosh Q. Optical properties of bismuth oxide nanoparticles synthesized by pulsed laser ablation in liquids. Science of Advanced Materials. 2012;**4**:1-4

[12] Geng J, Hou WH, Lv YN, Zhu JJ, Chen HY. One-dimensional $BiPO_4$ nanorods and two dimensional BiOCl lamellae: Fast low temperature sonochemical synthesis, characterization, and growth mechanism. Inorganic Chemistry. 2005;**44**;8503-8509

[13] Gipson K, Kucera C, Stadther D, Stevens K, Ballato J, Brown P. The influence of synthesis parameters on particle size and photoluminescence characteristics of ligand capped Tb^{3+}:LaF_3. Polymers. 2011;**3**:2039-2052

[14] Gutierrez M, Henglein A. Nanometer sized Bi particles in aqueous solution: Absorption spectrum and some chemical properties. The Journal of Physical Chemistry. 1996; **100**:7656-7661

[15] Hardcastle FD, Wachs IE. The molecular structure of bismuth oxide by Raman spectroscopy. Journal of Solid State Chemistry. 1992;**97**:319-331

[16] Haro-Poniatowski E, Serna R, de Castro MJ, Suarez- Garcia A, Afonso CN, Vickridge I. Size dependent thermo optical properties of embedded Bi nanostructures. Nanotechnology. 2008;**19**:485708 (6 p)

[17] Hunderi O. Optical properties of metallic calcium. Journal of Physics F: Metal Physics. 1976;**6**:1223-1229

[18] Karthikeyan B, Udayabhaskar R, Kishore A. Optical and phonon properties of Sm-doped α-Bi_2O_3 micro rods. Applied Physics A. 2014;**117**:1409-1414

[19] Kaur G, Rai SB. Luminescence properties of Sm, $Tb(Sal)_3Phen$ complex in polyvinyl alcohol: An approach for white-light emission. Journal of Physics D: Applied Physics. 2011;**44**:425306 (6 p)

[20] Kaur G, Dwivedi Y, Rai SB. Synthesis, structural, thermal and optical studies of rare earth coordinated complex: $Tb(Sal)_3Phen$. Materials Chemistry and Physics. 2011; **130**:1351-1356

[21] Kaur G, Dwivedi Y, Rai SB. Study of enhanced red emission from $Sm(Sal)_3Phen$ ternary complexes in Poly Vinyl Alcohol film. Optics Communication. 2010;**283**:3441-3447

[22] Kaur G, Verma RK, Rai DK, Rai SB. Plasmon-enhanced luminescence of Sm complex using silver nanoparticles in polyvinyl alcohol. Journal of Luminescence. 2012;**132**:1683-1687

[23] Kaur G. Rare Earths, Polymers and their Spectroscopy: Synthesis, Characterization, An Approach to Enhance Luminescence Intensity and Yield White light, LAP Lambert Academic Publications; 2016. 978-3-659-88306-4.

[24] Kaur G, Kumar B, Verma RK, Rai SB. Bismuth induced enhanced green emission from terbium ions and their complex in thin films. Dalton Transactions. 2014;**43**:11014-11018

[25] Kirkendall EO. Diffusion of zinc in alpha brass. Transactions of the American Institute of Mining, Metallurgical and Petroleum Engineers. 1942;**147**:104-109

[26] Kumar B, Kaur G, Singh P, Rai SB. Synthesis, structural, optical and electrical properties of metal nanoparticle–rare earth ion dispersed in polymer film. Applied Physics B. 2013;**110**:345-351

[27] Kumar B, Kaur G, Rai SB. Acetylsalicylic acid sensitized lasing luminescence of terbium complex in PVA: A case of energy avalanche via 1,10-phenanthroline. Journal of Photochemistry and Photobiology A: Chemistry. 2017;**332**:413-421

[28] Kumar B, Kaur G, Rai SB. Fluorescence quenching of Tb^{3+} ions with spherical Au NPs in PVA: Structural, thermal and optical investigations. Bulletin of Laser and Spectroscopic Society of India. 81-89

[29] Kumar B, Kaur G, Singh P, Rai SB. Anomalous electrical properties of polyvinyl alcohol films with Tb^{3+} ions and copper nanoparticles in different solvents. RSC Advances. 2015;**5**:1648-1654

[30] Kumari L, Lin JH, Ma YR. One dimensional Bi_2O_3 nanohooks: Synthesis, characterization and optical properties. Journal of Physics: Condensed Matter. 2007;**19**:406204 (11 p)

[31] Narang SN, Patel ND, Kartha VB. Infrared and Raman spectral studies and normal modes of α-Bi_2O_3. Journal of Molecular Structure. 1994;**327**:221-235

[32] Perez AJS, Lopez MAC, Luckie RAM, Mendieta VS, Nunez UF, Alatorre AJ. Structural evolution of Bi_2O_3 prepared by thermal oxidation of bismuth nano-particles. 2005;**18**:4-8

[33] Peng M, Zollfrank C, Wondraczek L. Origin of broad NIR photoluminescence in Bismuthate glass and Bi doped glasses at room temperature. Journal of Physics: Condensed Matter. 2009;**21**:285106 (6 p)

[34] Reisfeld R, Saraidarov T. Innovative materials based on sol-gel technology. Optical Materials. 2006;**28**:64-70

[35] Reisfeld R, Pietraszkiewicz M, Saraidarov T, Levchenko V. Luminescence intensification of lanthanide complexes by silver nanoparticles incorporated in sol-gel matrix. Journal of Rare Earths. 2009;**27**;544-549

[36] Serna R, De Sande JCG, Ballesteros JM, Afonso CN. Spectroscopic ellipsometry of composite thin films with embedded Bi nanocrystals. Journal of Applied Physics. 1998;**84**:4509

[37] Sivaramakrishnan S, Muthukumar VS, Sivasankara Sai S, Venkataramaniah K, Reppert J, Rao AM, Anija M, Philip R, Kuthirummal N. Nonlinear optical scattering and absorption in bismuth nanorod suspensions. Applied Physics Letters. 2007;**91**:093104

[38] Silversmith AJ, Boye DM, Brewer KS, Gillespie CE, Lu Y, Campbell DL. $^5D_3$-^7F_J emission in terbium-doped sol-gel glasses. Journal of Luminescence. 2006;**121**:14-20

[39] Talapin DV, Lee JS, Kovalenko MV, Shevchenko EV. Prospects of colloidal nanocrystals for electronic and optoelectronic applications. Chemical Reviews. 2010;**110**:389-458

[40] Toudert J, Serna R, de Castro MJ. Exploring the optical potential of nano-bismuth: Tunable surface plasmon resonances in the near ultraviolet-to-near infrared range. The Journal of Physical Chemistry C. 2012;**116**:20530-20539

[41] Wang YW, Hong BH, Kim KS. Size control of semimetal bismuth nanoparticles and the UV-visible and IR absorption spectra. The Journal of Physical Chemistry B. 2005; **109**:7067-7072

[42] Wang Q, Song F, Lin S, Ming C, Zhao H, Liu J, Zhang C, Pun EYB. Effect of silver nanoparticles with different shapes on luminescence of samarium complex at two different excitation wavelengths. Journal of Nanoparticle Research. 2011;**13**:3861-3865

[43] Wang C, Shao C, Liu Y, Zhang L. Photocatalytic properties BiOCl and Bi_2O_3 nanofibers prepared by electrospinning. Scripta Materialia. 2008a;**59**:332-335

[44] Wang F, Tang R, Yu H, Gibbons PC, Buhro WE. Size and shape-controlled synthesis of bismuth nanoparticles. Chemistry of Materials. 2008b;**20**:3656-3662

[45] Yu J, Wei B, Zhu L, Gao H, Sun W, Xu L. Flowerlike C-doped BiOCl nanostructures: Facile wet chemical fabrication and enhanced UV photocatalytic properties. Applied Surface Science. 2013;**284**:497-502

[46] Zhao Y, Zhang Z, Dang H. A simple way to prepare bismuth nanoparticles. Materials Letters. 2004;**58**:790-793

[47] Zhang L, Wang W, Sun S, Sun Y, Gao E, Xu J. Water splitting from dye wastewater: A case study of BiOCl/copper(II) phthalocyanine composite photocatalyst. Applied Catalysis B. 2013;**132-133**:315-320

[48] Kumar, et al. RSC Advances. 2016;**6**:26984-26992

Section 3

Acrylate Mitigation in Foods and Water

Chapter 7

Mitigation of Acrylamide in Foods: An African Perspective

Oluwafemi Ayodeji Adebo, Eugenie Kayitesi, Janet Adeyinka Adebiyi, Sefater Gbashi, Makumba Chewe Temba, Adeseye Lasekan, Judith Zanele Phoku and Patrick Berka Njobeh

Additional information is available at the end of the chapter

http://dx.doi.org/10.5772/intechopen.68982

Abstract

Acrylamide (ACR) is a possible human carcinogen, with neurotoxic properties. It is a heat-generated food toxicant particularly found in carbohydrate-rich foods. Its occurrence is of global concern and constitutes a major challenge to food safety, due to its presence in several thermally processed foods worldwide. Since its discovery, ACR has been recognized as one of the most widely investigated heat-induced food contaminant, and several reports on its formation and occurrence since its discovery have been reported. However, information on the extent of ACR occurrence in foods consumed in different parts of Africa is rather too limited. This is particularly a concern considering that most carbohydrate-based foods, subjected to varying degrees of thermal processing, are consumed as staple diets almost on daily basis in the continent. As such, African populations may be exposed to high levels of ACR daily. Thus, this chapter covers the formation, occurrence and health impact of ACR in foods. It further summarizes previous studies looking at ACR reduction and mitigation strategies, especially those that may be applicable in the continent. Adequate sensitization of the populace about the prevention of ACR as a food contaminant is essential to ensure the safety of heat-processed carbohydrate-rich foods in the continent.

Keywords: acrylamide, prevention, toxicity, heat-processed foods, Africa

1. Introduction

The prevalence of acrylamide (ACR) in ready-to-eat diets and its toxicological effects currently on humans is a public concern. The formation of this heat-generated toxic substance in foods, principally in carbohydrate-rich foods, was first reported by Tareke et al. [1] and has since been identified as a global challenge in the food industry. It has been classified as a potential occupational (Group 2A) carcinogen by the International Agency for Research on Cancer (IARC) and some US government agencies [2]. This is due to the fact that ACR is known to potentially exhibit carcinogenic effects in experimental animals, albeit its dietary link to human cancer. Its neurotoxicity in humans is well known from accidental and occupational exposures and experimental studies in animals which have shown genotoxic, reproductive and carcinogenic effects [3].

For over a decade since its discovery, several studies have been published in the literature on its formation, presence in various food products and toxicity in different parts of the world [1, 4–6]. In contrast, there is a dearth of information on its incidence and prevalence in Africa. This is probably why no information regarding limits regulating ACR in foods has not been established or enforced. Although other regions of the world can be affected by ACR contamination, it could be easily identified that Africa can be the most affected. It can, however, be difficult to affirm this assertion considering the lack of well-established or insufficient data on ACR levels in processed foods, degree of human exposure and risk assessment in the continent. This chapter appraises studies presenting information on the formation of ACR in foods and toxicity associated with it in other parts of the world. The main strategies for controlling or preventing its occurrence in the literature are also reviewed herein with a view of their possible adoption in Africa.

2. Discovery and properties of acrylamide

Acrylamide (IUPAC name—prop-2-enamide) (**Figure 1**) was accidentally discovered in foods in April 2002 by a group of researchers in Sweden working on heat processing technology of carbohydrate-rich foods [1, 7]. It is a white, odourless and crystalline compound with the chemical formula C_3H_5NO, molar mass of 71.08 g/mol, melting point of 84.5°C, vapour pressure of 0.007 mmHg at 25°C and boiling point at 136°C [8]. ACR is soluble in chloroform, ether, ethanol and water and decomposes in the presence of acids, oxidizing agents, bases, iron and iron salts [9] to form ammonia, carbon dioxide, carbon monoxide and oxides of nitrogen [10]. ACR is a heat-induced contaminant naturally formed during industrial processing and home cooking of many foods daily consumed around the world [8, 11].

ACR is used as a chemical intermediate in the production of polyacrylamides, which are used as a flocculating agent for sewage/wastewater treatment and other industrial applications

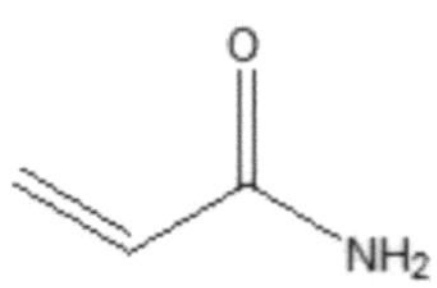

Figure 1. Chemical structure of acrylamide.

such as those in the formulation of several cosmetics [12]. Its application as a grouting agent is also extended to the construction of dam foundations, sewers and tunnels and cosmetics and in electrophoresis gels [13]. It has also been used in pesticide production, cement formulations, ore processing, sugar manufacturing, food packaging, plastic and paper production and for the prevention of soil erosion [2]. Sequel to its detection in foods and potential toxic effects, several studies have been initiated and reported worldwide, some of which will be highlighted in the succeeding sections of this chapter.

2.1. Formation of acrylamide

ACR is principally formed via Maillard reaction involving asparagine and carbonyl sources such as reducing sugars [14]. Although asparagine may be converted to ACR by thermally induced deamination and decarboxylation, carbohydrates are necessary to effect its conversion to ACR (**Figure 2**) [15]. While several other carbonyl compounds can enhance this reaction, α-hydroxyl carbonyl compounds such as glucose or fructose are more efficient [16]. Claus et al. [14] indicated that the first step in this reaction is the formation of a Schiff base intermediate as a low-energy alternative in decarboxylating this product intact. The formed Schiff base intermediate can either hydrolyze to form 3-aminopropionamide, a precursor of ACR, or further undergo elimination reaction leading to direct formation of ACR [17]. Nevertheless, the formation of ACR from reducing sugars and asparagine in the Maillard reaction represents the main formation route [14].

The formation of acrolein and acrylic acid through the dehydration of fats when heated at high temperature has been proposed as another mechanism of ACR formation [6, 14]. The studies of Becalski et al. [19] also indicated that ACR can be formed along with ammonia from

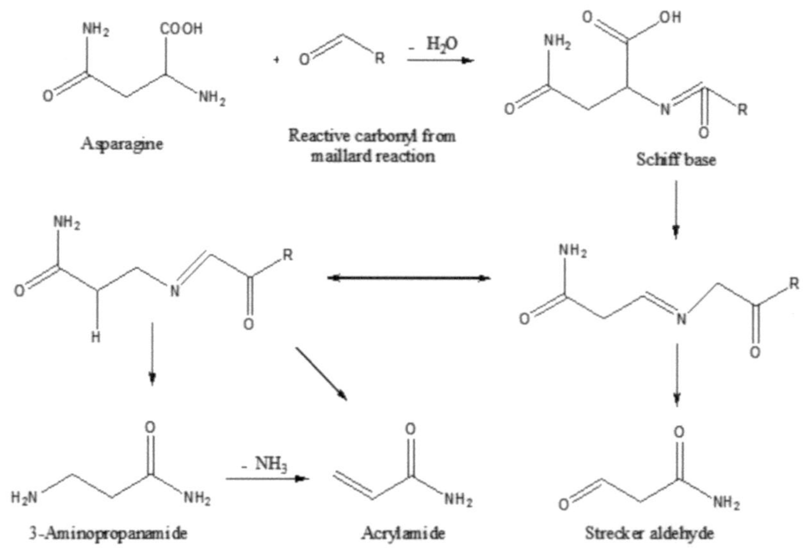

Figure 2. Proposed mechanism for the formation of the acrylamide in heat-treated foods. Adapted from Vleeschouwer et al. [18].

the degradation of amino acids. Although based on experiments with ammonium salts, acrolein and oils, this mechanism suggested might be inappropriate for ACR formation in foods [20]. Nevertheless, there are a number of factors that impact on the development of ACR in foods. Such factors are subsequently reviewed.

2.2. Factors affecting formation of acrylamide

2.2.1. *Processing conditions*

Food processing conditions such as time and temperature are vital factors affecting the formation and levels of ACR in food [21, 22]. However, the manner of heat transfer to foods (for instance, baking, frying, microwave heating and roasting) does not, however, necessarily impact the rate of ACR formation [23]. An exponential increase in ACR levels from 265 to 2130 μg/kg in French fries was observed when temperature increased from 150 to 190°C [21]. In potato slices with low and high surface-to-volume ratios (SVRs), ACR levels increased with increasing frying temperature and time, reaching maximum levels of 2500 and 18,000 μg/kg, respectively [24]. Studies on the effect of time and temperature on ACR formation in bread revealed that more ACR was formed in the crust layer and the levels increased with baking temperature and time [20, 25]. Roasting temperature and time had an impact on ACR formation in coffee beans [23, 26].

In Africa, traditional food processing operations and techniques are commonly applied but vary among different ethnic groups, communities and settlements. Heat processing operations such as frying, roasting and baking are common processes used in preserving and processing foods for further use. Thus, this presents a significant risk of ACR exposure. Elsewhere, it has been observed that concentrations of ACR are highly correlated with the degree of crust surface browning of processed foods [20]. These authors asserted that because ACR and brown colour of foods are formed due to Maillard reaction, it is most likely that ACR is formed in parallel with browning. Thus, the degree of surface browning (though may not necessarily indicate amount of ACR) could be used as a visual indicator of ACR formation in foods during cooking [25].

2.2.2. *Substrate composition*

The formation of ACR intermediates is usually determined by the amount and form of amino acids and sugars present [27]. It has thus been postulated that ACR formation and consequent concentration are relative to amino acid and sugar composition in the substrate [21, 27]. While these precursors affect ACR formation, the presence of other compounds that compete with amino acids and reducing sugars in the Maillard reaction are also vital compositional factors [21]. In potato and cereal products, ACR levels are highly correlated with glucose/fructose and asparagine concentrations [28]. Varieties in crop cultivar could also affect the reducing sugar content in the produce [29]. Short-term storage of potatoes at 4°C significantly increased the potential for ACR formation [30, 31]. Cooling potatoes to temperatures less than 10°C causes reducing sugars to increase, thereby increasing the potential for ACR formation [21, 30, 31]. Temperature and moisture levels in the food substrate are also other factors that affect ACR formation. While temperature has been discussed in the preceding section, Matthaus et al. [32] reported that a quick reduction of the water content in the outer layers of the product (as a consequence of high temperatures) favours ACR formation.

2.2.3. Soil properties and fertilization

Both nitrogen and sulphur are important compositions of the soil, and subsequent concentrations and amount have significant effect on the formation of ACR precursors [33–36]. According to Halford et al. [34], increasing soil sulphur levels and reducing nitrogen levels can effectively decrease the levels of ACR precursors, such as asparagine [34]. In a study conducted by De Wilde et al. [33] on the influence of soil enrichment on ACR formation in potatoes, differences occurred in ACR formation from crops grown on different soils. The effect of nitrogen fertilizer usage on crops has been reported in the literature to have an impact on asparagine and reducing sugar concentrations [37]. A decrease in nitrogen fertilizer resulted in a 30–65% increase in ACR formation [33, 38]. Moderate nitrogen fertilizer with a good provision of potassium may result in low levels of free asparagine and reducing sugars in tubers [39]. This clearly indicates that mineral composition due to either fertilizer employed or soil composition may impact on the presence and concentrations of ACR precursors. Deficiencies in phosphorus, potassium and magnesium can cause asparagine levels to rise in potato and wheat [34]. An appropriate balance between levels of fertilizer application and minimum requirements of the crop while taking into account possible environmental impacts and legal fertilizer limits should be reached to obtain food products less prone to ACR formation [33].

2.3. Occurrence and levels of acrylamide in foods

There have been considerable efforts made since the discovery of ACR in foods by regulatory agencies such as the US Food and Drug Administration (FDA), the World Health Organization (WHO) and the European Commission (EC) to gather data on food levels of ACR [40, 41]. However, none of these were reported from the continent of Africa. Prevalent sources of ACR differ among countries due to differences in the diet, method of preparing foods and the nature of soil/agricultural practices [42]. Cereal-, tuber- and coffee-related products contribute mostly to the sources of ACR intake [43]. Since ACR is present in a variety of food products which are consumed on a daily basis, the risks of exposure apply to almost all consumers. Children may be more vulnerable due to their smaller body mass as compared to adults [44].

Though a summary of reports on the incidence of ACR have been reported in other parts of the world [40, 41, 43, 44], a search of literature revealed few reports from Africa. The different studies reported are summarized in **Table 1** on ACR incidence levels reaching 12626, 9499, 7310 and 6968 μg/kg in South America, Africa, the Middle East and Europe, respectively. Of the very few incidences reported so far, only a few reported on African commodities were in Ghana [45], Kenya [46] and South Sudan (**Table 1**) [47, 48]. Though majority of these commodities were mainly baked and fried, this does not suggest that other heat-intense processed foods do not lead to ACR. The death of dogs after consuming the burnt part of maize meal is worth noting [49]. The death of these dogs was ascribed to ACR poisoning by the authors and should awaken intensive research on this. Further considering the fact that maize meal is a

Product/produce	Region of the world	Acrylamide content	Reference
Arabica	Europe	48–3210	[50]
Baby biscuit	Europe	588	[51]
Baby bread—rusks	Europe	660	[51]
Baby food	Europe	0–130	[41]
Baked *gorasa*	Africa	20	[48]
Baked *minnan*	Africa	17	[48]
Baked *hilmur*	Africa	59	[48]
Baked potato chips	Middle East	329–7310	[52]
Biscuit	Asia	119	[53]
		487	[54]
		232	[55]
	Europe	4200	[56]
		1177	[51]
		214	[57]
		1514	[58]
	South America	3180	[59]
Bread	Asia	103	[55]
	Middle East	90–802[a]	[60]
	Europe	2430	[56]
		695	[51]
Bread and rolls	Europe	400	[58]
Breakfast cereal	Asia	117.3	[53]
	Europe	1600	[56]
		762	[51]
		674	[58]
	South America	2288	[59]
Cakes	Europe	13–50	[61]
Candy bars	Europe	39–61	[61]
Cantonese moon cake	Asia	207	[62]
Cereal-based baby foods	Europe	353	[56]
		11–16	[61]
Cereals	Europe	52–1057	[41]
Cereal bar	Europe	820	[58]
Chocolate	Europe	750	[58]

Product/produce	Region of the world	Acrylamide content	Reference
Chocolate powder	South America	1017	[59]
Chocolate products	Asia	96.7	[53]
Coffee	Asia	7–19	[63]
		100–668	[64]
	Europe	1158	[56]
		16–503	[65]
		3800	[58]
	South America	3797	[59]
Coffee beans	Europe	172	[66]
		48–6968	[50]
Coffee substitute	Europe	5400	[58]
Cookies	Asia	50–700	[64]
Corn-based chips	Middle East	329–6360	[52]
	South America	78–441	[59]
Crackers	Europe	2666	[51]
	South America	194–1271	[59]
Crisp *mahua*	Asia	218	[62]
Fast food	Europe	210–2680	[50]
Follow-on formula	Europe	32–312	[67]
Follow-on formula (ready to eat)	Europe	4–46	[67]
French fries	Europe	20–1325	[41]
		320	[68]
		724[c]	[69]
	Asia	135	[53]
		78–496	[63]
		940	[55]
		441	[70]
French fries (home made)	Europe	2668	[56]
French fries (fast food rest.)		210–2922	[50]
	South America	12626–12661	[59]
Fried creole	South America	83–209	[59]
Fried chicken rolls	Asia	752	[55]
Fried eggplant	Africa	338	[47]
		325	[48]

Product/produce	Region of the world	Acrylamide content	Reference
Fried instant noodle	Asia	54	[54]
Fried potato	Africa	750	[47]
		227	[48]
Fried potato chips	Middle East	375–7024	[52]
Fried puffs	Asia	524	[55]
Fried sweet potato	Africa	1043	[45]
Fried *taamia*	Africa	68	[47]
Fries	Europe	3300	[58]
Gingerbread	Europe	2100	[58]
Home-cooked potato products	Europe	2175	[56]
Hot beverages	Europe	93–5399	[41]
Infant biscuits	Europe	3–516	[67]
Infant cereals in powder	Europe	65–296	[67]
Infant cereals (ready to eat)		11–52[a]	
Infant cereal with follow-on formula in powder		17–260	
Infant cereal with follow-on formula (ready to eat)		3–46	
Infant powdered formula	South America	1821	[59]
Instant cereal-based baby food	Europe	19.2–34.7	[61]
Jarred baby foods	Europe	162	[56]
		2–162	[67]
Juice	Europe	267	[41]
Mahua	Asia	234	[62]
Moon cake	Asia	201	[62]
Non-fried instant noodles	Asia	5	[54]
Nuts products	Asia	105	[53]
Paicha	Asia	214	[62]
Plantain chips	Africa	568	[45]
Popcorn	Europe	1100	[58]
	South America	781	[59]
Potato	Europe	131–5360	[78]
Potato-based chips	Middle East	375–7310	[52]
Potato chips	Asia	1021	[54]

Product/produce	Region of the world	Acrylamide content	Reference
		151	[53]
		0.4–14	[63]
		330–2300	[64]
		723	[70]
		233	[71]
	Europe	18–1782	[50]
	Middle East	90–800[a]	[60]
	South America	82–1852	[59]
Potato crisps	Africa	4565	[72]
		ND[b]–9499	[46]
	Asia	244–1688	[73]
	Europe	30–2300	[1]
		59–2336	[74]
		4180	[56]
		2311[c]	[75]
		954[c]	[69]
		3200	[58]
	South America	40–1770	[76]
Powdered baby food	Europe	174	[51]
Robusta	Europe	160–6968	[50]
Seasoned laver	Asia	103	[53]
Soft bread	South America	102–594	[59]
Sweet *binggan*	Asia	226	[62]
Taco, tostada and tortilla products	Europe	29–794	[41]
Tajadas	Europe	240	[77]
Tea products	Asia	108	[53]
Toast	Asia	530	[71]
	Europe	460	[58]
Twisted cruller	Asia	209	[71]
Wafer	South America	687–2497	[59]
Yougao	Asia	212	[62]
Youtiao	Asia	248	[62]

[a]µg/L.
[b]ND, not detected.
[c]Average value.

Table 1. Reported occurrence of acrylamide in foods by regions.

staple food in Southern African and gets burnt during its preparation, there are indications of a huge risk of ACR exposure to millions of individuals consuming this product daily.

2.4. Toxicity of acrylamide

The neurotoxicity of ACR in humans is well known from occupational and accidental exposures [79]. Owing to its low molecular weight and polarity, ACR is readily distributed and incorporated in mammals [80]. After ingestion, ACR is rapidly circulated throughout the whole body via the bloodstream [81] and can be found in the liver, kidney, thymus, brain, heart and human breast milk [82]. The conjugation of ACR to glutathione, and its epoxidation to glycidamide in the liver via cytochrome P450, is one of the major metabolic routes [83]. The formation of glycidamide is considered to be the critical step for the toxic effects of ACR and its metabolites. ACR and glycidamide, the latter at a much higher rate, can react with macromolecules such as haemoglobin and enzymes [80]. According to the European Food Safety Authority, ACR and its metabolite glycidamide have shown evidences of genotoxicity (DNA damage) and carcinogenicity [44]. Although evidence from studies on human exposure and possible causes of cancer is currently limited, epidemiological studies designed to target different populations and different organs in relation to cancer risks have been presented, with absolutely none reported from the African continent [84].

Calleman [85] reported peripheral neuropathy symptoms of highly exposed workers in China. Characterized by numbness of hands and feet, ataxia and skeletal muscle weakness, ACR has been shown to be toxic to both the central and peripheral nervous system [86]. ACR induces nerve terminal degeneration [79] and has deleterious effects on the thalamus, hippocampus and cerebral cortex [79, 86]. A recent study demonstrated evidence of ACR neurotoxic effects of fried potato chips on rat postnatal development, causing cerebellar cortical defects and myodegeneration of the gastrocnemius muscle during the postnatal development of pups [87]. It has been postulated that neurotoxicity of ACR might be cumulative as the same neurotoxic effects can be seen at low and high doses of ACR with the low doses requiring longer exposure [86, 88].

In 1994, ACR was classified by the International Agency for Research on Cancer (IARC) as Group 2A, indicating that it is probably carcinogen to humans (Group 2A) [89, 90]. This was based on positive bioassay results in rodents, buttressed by evidence that ACR is transformed in mammalian tissues to a more reactive genotoxic metabolite (glycidamide) [8]. Evidence on experimental rodents indicates that ACR causes tumours in the skin, uterus, lungs, brain, thyroid and mammary gland [91]. The genotoxicity of ACR and glycidamide is also manifested as both clastogenicity and mutagenicity. ACR has proven to be genotoxic in vivo to the somatic and germ cells as well as to cell cultures [8] and mammalian cells [81]. As indicated by Rice [91], the oxidation of this contaminant to glycidamide is the prerequisite for genotoxicity of ACR. This is attributed to the higher reactivity of this metabolite (glycidamide) to form adducts with DNA [83].

For cancer-related effects, the margin of exposures (MOEs) of ACR have been estimated to range from 50 for high-consuming toddlers to 425 for average adult consumers. These numbers indicate concerns for public health [44]. Essentially, since any level of exposure to a genotoxic substance could possibly cause DNA damage and lead to cancer, no tolerable daily intake (TDI) of ACR is set by European scientists [44], not to mention Africa. Nonetheless, Shipp et al. [13] reported that ACR administered to drinking water of rodents at doses of ≥5 mg/kg bw/day resulted in significant decreases in the number of live pups. At higher doses, signs of copulatory behaviour as well as effects on sperm motility and morphology were observed by these authors. ACR toxicity in male animals includes decrease in sperm number/ abnormal sperm, decrease in fertility rates, degeneration of the epithelial cells of the seminiferous tubules and retarded development of pups [92]. These reproductive toxic effects may be attributed to the interfering effect of ACR on the kinesin motor proteins, resulting in a reduced sperm motility and subsequent fertilization [92].

3. Prevention and mitigation of acrylamide

Agencies such as the FAO and WHO in collaboration with the academia and food industry have put forth strategies for reducing levels of ACR in food. In Europe, food manufacturers have collaborated with researchers and the academia through the Confederation of the Food and Drink Industries of the EU (CIAA) to produce series of strategies called the 'CIAA Acrylamide Toolbox' for decreasing ACR levels in different foods [93]. However, it should be noted that designing mitigation strategies is quite challenging, considering the fact that precautions must be taken to avoid compromising the nutritional, chemical, physical and microbiological quality and safety of the food. Accordingly, such measures must not result into the formation of other process contaminants nor detrimentally affect the organoleptic properties and acceptability of the final product [20, 94].

3.1. Methods that interrupt reactions leading to acrylamide formation

Several approaches have been successful at preventing ACR formation by preventing the key reactions responsible for generating it. Lowering the pH of foods blocks the nucleophilic addition of asparagine with a carbonyl compound, preventing the formation of the Schiff base, a critical intermediate in the formation of ACR [95, 96]. While this approach could be successful in lowering ACR levels in fried potato products, it may bring about undesirable taste to foods [21]. The use of organic acids and the addition of mono- and divalent cations (Na^+ or Ca^{2+}) to foods are other approaches of mitigating ACR by preventing the Schiff base formation [95–97]. The addition of proteins or free amino acids other than asparagine has also been investigated as a strategy for reducing ACR formation by causing competitive reactions and/or covalently binding ACR via Michael addition reactions [95, 96]. These additions however had low-to-moderate success at decreasing ACR levels in both cereal-based and potato foods [98].

3.2. Treatments that reduce acrylamide precursor's levels

As asparagine and reducing sugars are the major ACR precursors in foods, eliminating either of these substrates is a viable way to reduce ACR formation [94]. Procedures for achieving this include rinsing and blanching treatments, using asparaginase, fermentation and controlling storage conditions [21].

Rinsing, blanching and soaking treatments have been effective at reducing ACR formation in potato products [4, 21]. Soaking potato slices in water before frying resulted in over 50% reduction in ACR [21]. Further experiments by blanching slices in warm or hot water removed more glucose and asparagine than ordinary water immersion [4]. Changing the design of frying units to reverse the flow direction of the heated oil may alter the thermal load, which will reduce ACR levels in finished products [99]. Blanching and soaking treatments reduce ACR formation by leaching out asparagine and sugars from the surface of the slices [4]. Using asparaginase, an enzyme which hydrolyzes asparagine into aspartic acid and ammonia, has successfully reduced ACR levels in potato and bakery products [4, 20]. Asparaginase treatment of gingerbread dough resulted in a 75% decrease in free asparagine and a 55% reduction in ACR levels in the baked products [20]. To this effect, two commercial asparaginase preparations have been developed and are available in the market: Acrylaway® (Novozymes, Denmark) and PreventAse™ (DSM Food Specialties, Denmark), respectively, synthesized from *Aspergillus oryzae* and *Aspergillus niger*. They are generally recognized as safe (GRAS) ingredients [100].

Likewise, fermentation with yeast has been identified as a way to reduce ACR through the elimination of free asparagine [28]. A 2 h fermentation of rye and whole wheat dough caused a 77 and 87% reduction in ACR levels in rye and grain breads, respectively [101]. Yeast fermentation was observed to be more effective than sourdough fermentation in reducing the asparagine content of the dough [101]. Ingredients and additives may also increase ACR formation during baking of cereal-based products. In a study by Amrein et al. [20], baking agent and ammonium bicarbonate reportedly improved ACR formation in bakery products, possibly by creating more reactive carbonyl compounds. Using an alternative baking agent (sodium hydrogen carbonate), sucrose rather than honey or inverted sugar syrup can also reduce ACR content by more than 60% [20].

3.3. Modifying processing/cooking conditions

A reduction in cooking temperatures and times can decrease ACR levels in foods. However, loss of desirable colour, flavour and texture may occur, since the Maillard reaction which is responsible for ACR formation also guarantees desirable flavour and colour compounds in heated food [28]. Conditions that minimize ACR in French fries involve optimizing frying or baking processes to obtain a surface golden in colour and crispy texture [21]. Blanching, soaking, parboiling and washing treatments may be adopted, as these can leach the reducing sugar/asparagine reactants before the subsequent cooking step [102]. Overall, prolonged baking/frying and excessive browning should be avoided to minimize ACR formation in baked and fried products. Since a linear relationship exists between ACR formation and baking process, there is a need to ensure proper and optimum cooking endpoint to minimize ACR formation. This suggests that the degree of surface browning could be used as a visual indicator of ACR formation during cooking.

3.4. Agronomic factors

Selective crop propagation is a potential strategy for controlling ACR levels by decreasing levels of ACR precursors [94]. Since the first occurrence of ACR in foods, several researchers have demonstrated the significance of variety and cultivar selection on the formation of ACR [103]. Amrein et al. [29] found a 50-fold variation in total reducing sugars in the different potato cultivars the authors studied. Cultivars with low reducing sugars were more suitable for potato products, cooked or processed at high temperatures [29]. Konings et al. [28] established the significant impact of fertilizer application rate on ACR levels mainly due to differences in crude protein and asparagine and contents. A study by Claus et al. [38] demonstrated the effect of nitrogen-based fertilizers in causing high amounts of protein and amino acid. This resulted in increased ACR levels in breads, ranging from 10.6 to 55.6 μg/kg [38]. Producers should best adopt effective fertilizer application regimes that will subsequently yield suitable produce for processing, as this influence the levels of reducing sugars [29]. Tubers should be harvested at full maturity as selection of immature tubers for further processing increases the chances of ACR occurrence, because they have relatively higher reducing sugars and produce products with potentially higher ACR levels [104, 105]. Unfortunately, most African subsistence farmers harvest immature tubers to immediately obtain income for their needs. While this should be discouraged, effective handling, packaging and storage of produce must also be emphasized. Selection and use of crop varieties that are low in ACR precursors will most definitely help reduce ACR occurrence. Storage is also another component and practicable way of mitigating ACR in foods. While storage of potatoes at low temperatures is generally meant to minimize shrinkage and spoilage, studies have shown that low temperatures tend to increase sugar levels (an ACR precursor) [106]. Though for a short-term storage, hot temperature is desirable (this can however lead to sprouting, which can be controlled using suppressants); for long term, a minimum storage temperature of 6°C is desirable [97].

3.5. Antioxidants and other phytochemicals

According to Kahkeshani et al. [107], the correlation between antioxidants and ACR can be considered from two different points of view, namely, antioxidants as exogenous additives and as endogenous secondary metabolites. According to these authors, lack of sufficient studies and discord in the results available from literature hinders a logical judgement about the effectiveness of phytochemicals against ACR. While some reports have reported their beneficial effect, some have been shown to facilitate ACR production [107]. While these compounds can react with asparagine to produce ACR, they could also possibly react with the amide group of the intermediates in Maillard reaction and block ACR formation [107–110]. The oxidation of polyphenols to corresponding quinones, which can react with 3-animopropionamide (3-APA), thus preventing the deamination of 3-APA to ACR, has also been proposed as a mechanism for ACR reduction by these compounds [111]. Nonetheless, studies demonstrating the effectiveness of these compounds towards the reduction of ACR have been presented in the literature. Fernandez et al. [112] recorded a 50% ACR reduction after the addition of a flavonoid spice mix to potato chips. Zhang and Zhang [108] reported a 76% reduction in ACR after French fries were dipped into extracts of bamboo leaves with antioxidant properties, while a 59% decrease in ACR was recorded when fried chicken wings were dipped into same extracts [111]. Further reports in the literature on this have been adequately reviewed by Kahkeshani et al. [107] and can be consulted for further reading. Interestingly, Africa is the home to a vast and diverse number of plants and other botanicals with rich phytochemicals. Extracts of these plants and herbs are used for various purposes prominently in traditional medicines. Application of such extracts would go a long way in the mitigation of ACR occurrence in foods.

3.6. Genetic modification

As defined by Key et al. [113], genetically modified crops/plants are those that have been genetically altered through the use of recombinant DNA technology. This may be to express a gene not native to the plants or to modify endogenous ones [113]. This issue has in the recent years attracted worldwide attention especially regarding its risk to the environment and human health. While it is widely accepted in many parts of the world, including parts of America and Asia, it still remains a controversial issue in Europe [35]. Most African continents follow the latter, as concerns regarding the immediate and long-term effects of genetically modified crops are major hindrances to adopting this technology. Although different authors, government agencies and international organizations have backed and supported genetic modification, there is still a stiff opposition against its acceptance. Inconsistencies in free sugar, amino acid and asparagine contents in crops of different cultivars (varieties) and genotypes however suggest that the varying concentration of these parameters is due to genetic variations [19, 29, 38, 114]. Consequently, fast tracking the natural breeding process through the use of genetic engineering to develop cultivars (varieties) with lower concentration of asparagine and reducing sugars should be possible and encouraged [35]. It has been shown that simultaneous silencing of the genes (StAst1 and StAst2) that encode for asparagine synthetase which is the enzyme that catalyse the formation of asparagine in potato, significantly reduced the levels of asparagine in the transgenic crop [115]. Another study demonstrated reduction in the concentration of reducing sugars, which also participate in reaction leading to ACR formation. For instance, since reducing sugar is accumulated during the cold storage of potato (cold-induced sweetening), silencing of the enzyme acid invertase resulted in potato with reduced concentration of fructose and glucose as well as low ACR concentration when processed into French fries [116]. While genetic modification continues to be controversial, the farmers, the food industry and other vital stakeholders should be proactive in the development of crop varieties that would yield lower ACR levels in food.

4. Conclusion and future prospects

This chapter gives an overview of ACR in foods, significant progress in its formation and mitigation strategies with a dearth of information in Africa. Its occurrence and exposure in other parts of the world have been extensively reviewed by other authors with little focus on the African continent. Starch-based foods and food products constitute a major and basic daily diet for millions in the developing world, particularly in Africa. Coupled with the myriad of associated traditional heat processing operations, it is justified to conclude that inhabitants in this region are exposed to high risk of ACR contamination. This is expected to stimulate interest among scientists working in the field of food safety and quality, for making better efforts towards investigating the occurrence and exposure of ACR in Africa. With such data lacking, there is also insufficient information on the impact of lower levels of exposure to ACR content, which needs to be established. Concerted efforts must also be directed towards this, using validated models of predicting dose exposure and mechanism of toxicity relationship to assist in measuring the public health risk of ACR in foods. Furthermore, adequate enlightenment and sensitization of the populace by government agencies and the industry about the dangers and possible ways of reducing this food contaminant must be provided and emphasized.

Acknowledgements

The authors would like to acknowledge the financial support via the Global Excellence and Stature (GES) Fellowship of the University of Johannesburg (UJ) provided to the main author (O. A. Adebo). This work was also partly supported by the National Research Foundation (NRF) Center of Excellence (CoE) in Food Security co-hosted by the University of Pretoria (UP) and the University of Western Cape (UWC), South Africa.

Author details

Oluwafemi Ayodeji Adebo[1]*, Eugenie Kayitesi[1], Janet Adeyinka Adebiyi[1], Sefater Gbashi[1], Makumba Chewe Temba[1], Adeseye Lasekan[2], Judith Zanele Phoku[3] and Patrick Berka Njobeh[1]*

*Address all correspondence to: oluwafemiadebo@gmail.com and pnjobeh@uj.ac.za

1 Department of Biotechnology and Food Technology, Faculty of Science, University of Johannesburg, Doornfontein, Gauteng, South Africa

2 Department of Food Science and Human Nutrition, University of Maine, Orono, ME, USA

3 Water and Health Research Unit, Faculty of Health Science, University of Johannesburg, Doornfontein, Gauteng, South Africa

References

[1] Tareke E, Rydberg P, Karlsson P, Eriksson S, Tornqvist M. Analysis of acrylamide, a carcinogen formed in heated foodstuffs. Journal of Agricultural and Food Chemistry. 2002;**50**: 4998-5006. DOI: 10.1021/jf020302f

[2] Dotson GS. NIOSH Skin Notation (Sk) Profile: Acrylamide [CAS No. 79-06-1]. US Department of Health and Human Services, Public Health Service, Centre for Disease Control and Prevention, National Institute for Occupational Safety and Health DHHS (NIOSH) Publication No. 2011-139; 2011. pp. 1-13. Atlanta, USA.

[3] Bolger MP, Jean-Charles L, Setzer RW. Application of the Margin of Exposure (MoE) approach to substances in food that are genotoxic and carcinogenic. Example: Acrylamide (CAS No. 79-06-1). Food and Chemical Toxicology. 2010;**48**:23-33. DOI: 10.1016/j.fct.2009. 11.040

[4] Pedreschi F, Kaack K, Granby K. The effect of asparaginase on acrylamide formation in French fries. Food Chemistry. 2008;**109**:386-392. DOI: 10.1016/j.foodchem.2007.12.057

[5] Arvanitoyannis IS, Dionispoulou N. Acrylamide: Formation, occurrence in food products, detection methods and legislation. Critical Reviews in Food Science and Nutrition. 2014;**54**:708-733. DOI: 10.1080/10408398.2011.606378

[6] Krishnakumar T, Visvanathan R. Acrylamide in food products: A review. Journal of Food Processing and Technology. 2014;**5**:1-9. DOI: 10.4172/2157-7110.1000344

[7] Svensson K, Abramsson L, Becker W, Glynn A, Hellenas KE, Lind Y, Rosen J. Dietary intake of acrylamide in Sweden. Food and Chemical Toxicology. 2003;**41**:1581-1586. DOI: 10.1016/S0278-6915(03)00188-1

[8] Arribas-Lorenzo G, Morales FJ. Recent insights in acrylamide as carcinogen in foodstuffs. In: Fishbein JC, Heilman JM, editors. Advances in Molecular Toxicology. Elsevier: Amsterdam; 2012. pp. 163-184

[9] Friedman M. Chemistry, biochemistry and safety of acrylamide. A review. Journal of Agricultural and Food Chemistry. 2003;**51**:4504-4526. DOI: 10.1021/jf030204+

[10] Eriksson S. Acrylamide in food products: Identification, formation and analytical methodology, [PhD thesis], Stockholm, Sweden: Department of Environmental Chemistry, Stockholm University; 2005

[11] Mottram DS, Wedzicha BL, Dodson AT. Acrylamide is formed in the maillard reaction. Nature. 2002;**419**:448-449. DOI: 10.1038/419448a

[12] Touzé S, Guerin V, Guezennec AG, Binet S, Togola A. Dissemination of acrylamide monomer from polyacrylamide-based flocculant use—sand and gravel quarry case study. Environmental Science and Pollution Research International. 2015;**22**:6423-6430. DOI: 10.1007/s11356-014-3177-0

[13] Shipp A, Lawrence G, Gentry R, McDonald T, Bartow H, Bounds J. Acrylamide: Review of toxicity data and dose-response analyses for cancer and noncancer effects. Critical Reviews in Toxicology. 2006;**36**:481-608. DOI: 10.1080/10408440600851377

[14] Claus A, Carle R, Schieber A. Acrylamide in cereal products: A review. Journal of Cereal Science. 2008;**47**:118-133. DOI: 10.1016/j.jcs.2007.06.016

[15] Yaylayan VA, Wnorowski A, Perez-Locas C. Why asparagines needs carbohydrates to generate acrylamide. Journal of Agricultural and Food Chemistry. 2003;**51**:1753-1757. DOI: 10.1021/jf0261506

[16] Yaylayan V, Stadler R. Acrylamide formation in food: A mechanistic perspective. Journal of AOAC International. 2005;**88**:262-267

[17] Granvogl M, Jezussek M, Koehler P, Schieberle P. Quantitation of 3-aminopropionamide in potatoes—a minor but potent precursor in acrylamide formation. Journal of Agricultural and Food Chemistry. 2004;**52**:4751-4757. DOI: 10.1021/jf049581s

[18] Vleeschouwer K, Van der PI, Van LA, Hendrickx ME. Kinetics of acrylamide formation/elimination reactions as affected by water activity. Biotechnology Progress. 2007;**23**:722-728. DOI: 10.1021/bp060389f

[19] Becalski A, Lau BPY, Lewis D, Seaman SW, Hayward S, Sahagian M, Ramesh M, Leclerc Y. Acrylamide in French fries: Influence of free amino acids and sugars. Journal of Agricultural and Food Chemistry. 2004;**52**:3801-3806. DOI: 10.1021/jf0349376

[20] Amrein TM, Schonbachler B, Escher F, Amado R. Acrylamide in gingerbread: Critical factors for formation and possible ways for reduction. Journal of Agricultural and Food Chemistry. 2004;**52**:4282-4288. DOI: 10.1021/jf049648b

[21] Jackson LS, Al-Taher F. Effects of consumer food preparation on acrylamide formation. Advances in Experimental Medicine and Biology. 2005;**561**:447-465. DOI: 10.1007/0-387-24980-X_34

[22] Amrein TM, Andres L, Escher F, Amado R. Occurrence of acrylamide in selected foods and mitigation options. Food Additives and Contaminants. 2007;**24**:13-25. DOI: 10.1080/02652030701242558

[23] Stadler RH, Robert F, Riediker S. In-depth mechanistic study on the formation of acrylamide and other vinylogous compounds by the Maillard reaction. Journal of Agricultural and Food Chemistry. 2004;**52**:5550-5558. DOI: 10.1021/jf0495486

[24] Taubert D, Harlfinger S, Henkes L. Influence of processing parameters on acrylamide formation during frying of potatoes. Journal of Agricultural and Food Chemistry. 2004;**52**:2735-2739. DOI: 10.1021/jf035417d

[25] Surdyk N, Rosén J, Andersson R, Aman P. Effects of asparagine, fructose, and baking conditions on acrylamide content in yeast-leavened wheat bread. Journal of Agricultural and Food Chemistry. 2004;**52**:2047-2051. DOI: 10.1021/jf034999w

[26] Taeymans D, Wood J, Ashby P, Blank I, Studer A, Stadler RH, Gondé P, Van Eijck P, Lalljie S, Lingnert H, Lindblom M, Matissek R, Müller D, Tallmadge D, O'Brien J, Thompson S, Silvani D, Whitmore T. A review of acrylamide: An industry perspective on research, analysis, formation, and control. Critical Reviews Food Science and Nutrition. 2004;**44**:323-347. DOI: 10.1080/10408690490478082

[27] Ciesarova Z, Kiss E, Kolek E. Study of factors affecting acrylamide levels in model systems. Czech Journal of Food Science. 2006;**24**:133-137

[28] Konings EJM, Ashby P, Hamlet CG, Thompson GAK. Acrylamide in cereal and cereal products: A review on the progress in level reduction. Food Additives and Contaminants. 2007;**24**:47-59. DOI: 10.1080/02652030701242566

[29] Amrein TM, Bachmann S, Noti A, Biedermann M, Barbosa MF, Biedermann-Brem S, Grob K, Keiser A, Realini P, Escher F, Amado R. Potential of acrylamide formation, sugars, and free asparagine in potatoes: A comparison of cultivars and farming systems. Journal of Agricultural and Food Chemistry. 2003;**51**:5556-5560. DOI: 10.1021/jf034344v

[30] Biedermann M, Noti A, Biedermann-Brem S. Experiments on acrylamide formation and possibilities to decrease the potential of acrylamide formation in potatoes. Mitteilungenaus Lebensmitteluntersuchung und Hygiene. 2002;**93**:668-687

[31] Noti A, Biedermann-Brem S, Biedermann M. Storage of potatoes at low temperatures should be avoided to prevent increased acrylamide formation during frying or roasting. Mitteilungenaus Lebensmitteluntersuchung und Hygiene. 2003;**94**:167-180

[32] Matthaus B, Haase NU, Vosmann K. Factors affecting the concentration of acrylamide during deep-fat-frying of potatoes. European Journal of Lipid Science and Technology. 2004;**106**:793-801

[33] De Wilde T, De Meulenaer B, Mestdagh F. Influence of fertilization on acrylamide formation during frying of potatoes harvested in 2003. Journal of Agricultural and Food Chemistry. 2006;**54**:404-408. DOI: 10.1021/jf0521810

[34] Halford NG, Muttucumaru N, Curtis TY, Parry MA. Genetic and agronomic approaches to decreasing acrylamide precursors in crop plants. Food Additives and Contaminants. 2007;**24**:26-36. DOI: 10.1080/02652030701403093

[35] Halford NG, Curtis TY, Muttucumaru N, Postles J, Elmors JS, Mottram DS. The acrylamide problem: A plant and agronomic science issue. Journal of Experimental Botany. 2012;**63**:2841-2851. DOI: 10.1093/jxb/ers011

[36] Stojanovska S, Tomovska J. Factors influence to formation of acrylamide in food. Journal of Hygienic Engineering and Design. 2015;**13**:10-15

[37] Torres MDA, Parreño WC. Thermal processing and quality optimization. In: Singh J, Kaur L, editors. Advances in potato chemistry and technology. Elsevier: Amsterdam; 2009. pp. 163-219

[38] Claus A, Schreiter P, Weber A, Graeff S, Hermann W, Claupein W, Schieber A, Carle R. Influence of agronomic factors and extraction rate on the acrylamide contents in yeast leavened breads. Journal of Agricultural and Food Chemistry. 2006;**54**:8968-8976. DOI: 10.1021/jf061936f

[39] Heuser H, Gerendás J, Sattelmacher B. Einfluss der N- und k-dünung auf die gehalte an reduzierenden zuckern und freien aminosäuren. Kartoffel. 2005;**56**:308-313

[40] FDA (Food and Drug Administration). Survey Data on Acrylamide in Food: Individual Food Products. US Food and Drug Administration and Center for Food Safety and Applied Nutrition; 2006. Maryland, USA

[41] FDA (Food and Drug Administration). Survey Data on Acrylamide in Food: Total Diet Survey Results. US FDA and Center for Food Safety and Applied Nutrition (CFSAN); 2006. Maryland, USA

[42] Mills C, Mottram DS, Wedzicha BL. Acrylamide. In: Stadler RH, Lineback DR, editors. Process-Induced Food Toxicants. Hoboken, NJ: Wiley; 2009. pp. 23-50

[43] FDA (Food and Drug Administration). The 2006 Exposure Assessment for Acrylamide. US FDA and Center for Food Safety and Applied Nutrition (CFSAN); 2006. Maryland, USA

[44] EFSA (European Food Safety Authority). EFSA Explains Risk Assessment: Acrylamide in Food. EFSA Factsheet. Parma, Italy: European Food Safety Authority; 2015

[45] Quayson ET, Ayenor GS. Non-enzymatic browning and estimated acrylamide in roots, tubers and plantain products. Food Chemistry. 2007;**105**:1525-1529. DOI: 10.1016/j.foodchem.2007.05.066

[46] Ogolla JA, Abong GO, Okoth MW, Kabira JN, Imungi JK, Karanja PN. Levels of acrylamide in commercial potato crisps sold in Nairobi county, Kenya. Journal of Food and Nutrition Research. 2015;**3**:495-501. DOI: 10.12691/jfnr-3-8-4

[47] Omar MMA, Wan Ibrahim WA, Elbashir AA. Sol–gel hybrid methyltrimethoxysilane–tetraethoxysilane as a new dispersive solid-phase extraction material for acrylamide determination in food with direct gas chromatography–mass spectrometry analysis. Food Chemistry. 2014;**158**:302-309. DOI: 10.1016/j.foodchem.2014.02.045

[48] Omar MMA, Elbashir AA, Schmitz OJ. Determination of acrylamide in Sudanese food by high performance liquid chromatography coupled with LTQ Orbitrap mass spectrometry. Food Chemistry. 2015;**176**:342-349. DOI: 10.1016/j.foodchem.2014.12.091

[49] Le Roux-Pullen L, Lessing D Should veterinarians consider acrylamide that potentially occurs in starch-rich foodstuffs as a neurotoxin in dogs? Journal of the South African Veterinary Association. 2011;**82**:129-130

[50] Oroian M, Amariei S, Gutt G. Acrylamide in Romanian food using HPLC-UV and a health riskassessment.FoodAdditivesandContaminants.2015;**8**:136-141.DOI:10.1080/19393210.2015.1010240

[51] Cengiz MF, Gundiz CPB. Acrylamide exposure among Turkish toddlers from selected cereal-based baby food samples. Food and Chemical Toxicology. 2013;**60**:514-519. DOI: 10.1016/j.fct.2013.08.018

[52] Hariri E, Abboud MI, Demirdjian S, Korfal S, Mroueh M, Taleb RI. Carcinogenic and neurotoxic risks of acrylamide and heavy metals from potato and corn chips consumed by the Lebanese population. Journal of Food Composition and Analysis. 2015;**42**:91-97. DOI: 10.1016/j.jfca.2015.03.009

[53] Lee S, Yoo M, Koo M, Kim HJ, Kim M, Park SK, Shin D. In-house–validated liquid chromatography–tandem mass spectrometry (LC-MS/MS) method for survey of acrylamide in various processed foods from Korean market. Food Science and Nutrition. 2013;**1**:402-407. DOI: 10.1002/fsn3.56

[54] Zhu Y, Li G, Duan Y, Chen S, Zhang C, Li Y. Application of the standard addition method for the determination of acrylamide in heat-processed starchy foods by gas chromatography with electron capture detector. Food Chemistry. 2009;**109**:899-908. DOI: 10.1016/j.foodchem.2008.01.020

[55] Liu C, Luo F, Chen D, Qiu B, Tang X, Ke H, Chen X. Fluorescence determination of acrylamide in heat-processed foods. Talanta. 2014;**123**:95-100. DOI: 10.1016/j.talanta.2014.01.019

[56] EFSA (European Food Safety Authority). Scientific report of EFSA prepared by data collection and exposure unit (DATEX) on monitoring of acrylamide levels in food. The EFSA Scientific Report. 2009;**285**:1-26

[57] Vaclavik L, Capuano E, Gokmen V, Hajslova J. Prediction of acrylamide formation in biscuits based on fingerprint data generated by ambient ionization mass spectrometry employing direct analysis in real time (DART) ion source. Food Chemistry. 2015;**173**: 290-297. DOI: 10.1016/j.foodchem.2014.09.151

[58] Claeys W, De Meulenaer B, Huyghebaert A, Scippo ML, Hoet P, Matthys C. Reassessment of the acrylamide risk: Belgium as a case-study. Food Control. 2016;**59**:628-635. DOI: 10.1016/j.foodcont.2015.06.051

[59] Pacetti D, Gil E, Frega NG, Alvarez L, Duenas P, Garzon A, Lucci P. Acrylamide levels in selected Colombian foods. Food Additives and Contaminants. 2015;**8**:99-105. DOI: 10.1080/19393210.2014.995236

[60] Albishri HM, El-Hady DA. Eco-friendly ionic liquid based ultrasonic assisted selective extraction coupled with a simple liquid chromatography for the reliable determination of acrylamide in food samples. Talanta. 2014;**118**:129-136. DOI: 10.1016/j.talanta.2013.10.015

[61] Michalak J, Gujska E, Kuncewicz A. RP-HPLC-DAD studies on acrylamide in cereal-based baby foods. Journal of Food Composition and Analysis. 2013;**32**:68-73. DOI: 10.1016/j.jfca.2013.08.006

[62] Wang H, Feng F, Guo Y, Shuang S, Choi MMF. HPLC-UV quantitative analysis of acrylamide in baked and deep-fried Chinese foods. Journal of Food Composition and Analysis. 2013;**31**:7-11. DOI: 10.1016/j.jfca.2013.02.006

[63] Lim HH, Shin HS. A new derivatization approach with d-cysteine for the sensitive and simple analysis of acrylamide in foods by liquid chromatography–tandem mass spectrometry. Journal of Chromatography A. 2014;**1361**:117-124. DOI: 10.1016/j.chroma.2014.07.094

[64] Wu J, Shen YD, Lei HT, Sun YM, Yang JY, Xiao ZL, Wang H, Xu ZL. Hapten synthesis and development of a competitive indirect enzyme-linked immunosorbent assay for acrylamide in food samples. Journal of Agricultural and Food Chemistry. 2014;**62**:7078-7084. DOI: 10.1021/jf5015395

[65] Pugajeva I, Jaunbergs J, Bartkevics V. Development of a sensitive method for the determination of acrylamide in coffee using high-performance liquid chromatography coupled to a hybrid quadrupole Orbitrap mass spectrometer. Food Additives and Contaminants. 2015;**32**:170-179. DOI: 10.1080/19440049.2014

[66] Budryn G, Nebesny E, Oracz, J. Correlation between the stability of chlorogenic acids, antioxidant activity and acrylamide content in coffee beans roasted in different conditions. International Journal of Food Properties. 2015;**18**:290-302. DOI: 10.1080/10942912.2013.805769

[67] Mojska H, Gielecinska I, Stos K. Determination of acrylamide level in commercial baby foods and an assessment of infant dietary exposure. Food and Chemical Toxicology. 2012;**50**:2722-2728. DOI: 10.1016/j.fct.2012.05.023

[68] EFSA (European Food Safety Authority). Results on acrylamide levels in food from monitoring year 2008. EFSA Journal. 2010;**8**:1-31

[69] Sirot V, Hommet F, Tard A, Leblanc JC. Dietary acrylamide exposure of the French population: Results of the second French total diet study. Food and Chemical Toxicology. 2012;**50**:889-894. DOI: 10.1016/j.fct.2011.12.033

[70] Weijun Y. Direct determination of acrylamide in food by gas chromatography with nitrogen chemiluminescence detection. Journal of Separation Science. 2015;**38**:2272-2277. DOI: 10.1002/jssc.201500060

[71] Zhao H, Li N, Li J, Qiao X, Xu Z. Preparation and application of chitosan-grafted multiwalled carbon nanotubes in matrix solid-phase dispersion extraction for determination of trace acrylamide in foods through high-performance liquid chromatography. Food Analytical Methods. 2015;**8**:1363-1371. DOI: 10.1007/s12161-014-0022-5

[72] Mestdagh F, Lachat C, Katleen Baert K, Moons E, Kolsteren P, Van Peteghem C, Bruno De Meulenaer B. Importance of a canteen lunch on the dietary intake of acrylamide. Molecular Nutrition and Food Research. 2007;**51**:509-516. DOI: 10.1002/mnfr.200600253

[73] Boroushaki MT, Nikkhah E, Kazemi A, Oskooei M, Raters M. Determination of acrylamide level in popular Iranian brands of potato and corn products. Food and Chemical Toxicology. 2010;**48**:2581-2584. DOI: 10.1016/j.fct.2010.06.011

[74] Olmez H, Tuncay F, Ozcan N, Demirel S. A survey of acrylamide levels in foods from the Turkish market. Journal of Food Composition and Analysis. 2008;**21**:564-568. DOI: 10.1016/j.jfca.2008.04.011

[75] Douny C, Widart J, Maghuin-Rogister G, De Pauw, Scippo ML. Quantification of acrylamide in various Belgian potato products using solid phase extraction and liquid chromatography tandem mass spectrometry detection. Food and Public Health. 2012;**2**:137-141. DOI: 10.5923/j.fph.20120205.04

[76] Pedreschi P, Granby B, Risum J. Acrylamide mitigation in potato chips by using nacl. Food and Bioprocess Technology. 2010;**3**:917-921. DOI: 10.1007/s11947-010-0349-x

[77] Bassama J, Brat P, Boulanger R, Gunata Z, Bohuon P. Modeling deep-fat frying for control of acrylamide reaction in plantain. Journal of Food Engineering. 2012;**113**:156-166. DOI: 10.1016/j.jfoodeng.2012.04.004

[78] Elmore JS, Briddon A, Dodson AT, Muttucumaru N, Halford NG, Mottram DS. Acrylamide in potato crisps prepared from 20 UK-grown varieties: Effects of variety and tuber storage time. Food Chemistry. 2015;**182**:1-8. DOI: 10.1016/j.foodchem.2015.02.103

[79] FAO/WHO (Food and Agriculture Organization /World Health Organization). Summary and Conclusions of the Sixty-Fourth Meeting of the Joint FAO/WHO Expert Committee on Food Additives (JECFA). ftp://ftp.fao.org/es/esn/jecfa/jecfa64_summary.pdf; 2005

[80] Wilson KM, Rimm EB, Thompson KM, Mucci LA. Dietary acrylamide and cancer risk in humans: A review. Journal of Consumer Protection and Food Safety. 2006;**1**:19-27

[81] Abramsson-Zetterberg L. The dose-response relationship at very low doses of acrylamide is linear in the flow cytometer-based mouse micronucleus assay. Mutation Research. 2003;**535**:215-222. DOI: 10.1016/S1383-5718(02)00322-4

[82] Sorgel F, Weissenbacher R, Kinzig-Schippers M, Hofmann A, Illauer M, Skott A, Landersdorfer C. Acrylamide: Increased concentrations in homemade food and first evidence of its variable absorption from food, variable metabolism and placental and breast milk transfer in humans. Chemotheraphy. 2002;**48**:267-274. DOI: 69715

[83] Doerge DR, da Costa GG, McDaniel LP, Churchwell MI, Twaddle NC, Beland FA. DNA adducts derived from administration of acrylamide and glycidamide to mice and rats. Mutation Research. 2005;**580**:131-141. DOI: 10.1016/j.mrgentox.2004.10.013

[84] Capuano E, Fogliano V. Acrylamide and 5-hydroxymethyfurfural (HMF): A review on metabolism, toxicity, occurrence in food and mitigation strategies. LWT-Food Science and Technology. 2011;**44**:793-810. DOI: 10.1016/j.lwt.2010.11.002

[85] Calleman CJ. The metabolism and pharmacokinetics of acrylamide: Implications for mechanisms of toxicity and human risk estimation. Drug Metabolism Review. 1996;**28**:527-590. DOI: 10.3109/03602539608994018

[86] LoPachin RM. The changing view of acrylamide neurotoxicity. Neurotoxicology. 2004;**25**: 617-630. DOI: 10.1016/j.neuro.2004.01.004

[87] El-Sayya HI, El-Gammal HL, Habak LA, Abdel-Galil HM, Fernando A, Gaur RL, Ouhtit A. Structural and ultrastructural evidence of neurotoxic effects of fried potato chips on rat postnatal development. Nutrition. 2011;**27**:1066-1075. DOI: 10.1016/j.nut.2011.06.008

[88] Sickles DW, Sperry AO, Testino A, Friedman M. Acrylamide effects on kinesin related proteins of the mitotic/meiotic spindle. Toxicology and Applied Pharmacology. 2007;**222**: 111-121. DOI: 10.1016/j.taap.2007.04.006

[89] IARC (International Agency for Research on Cancer). Some industrial chemicals. IARC Monographs, Evaluation for Carcinogenic Risk of Chemicals to Humans. 1994;**60**:389-433

[90] NTP (National Toxicology Program). Report on Carcinogens, 12th ed; 2011. pp. 25-28

[91] Rice JM. The carcinogenicity of acrylamide. Mutation Research. 2005;**580**:3-20. DOI: 10. 1016/j.mrgentox.2004.09.008

[92] Tyl RW, Marr MC, Myers CB, Ross WP, Friedman MA. Relationship between acrylamide reproductive and neurotoxicity in male rats. Reproductive Toxicology. 2000;**14**:147-157

[93] CIAA. The CIAA Acrylamide 'Toolbox', Confederation of Food and Drink Industries of the EU. http://www.ciaa.be/documents/brochures/CIAA_Acrylamide_Toolbox_Oct2006. pdf; 2006

[94] CAST (Council for Agricultural Science and Technology). Acrylamide in food. CAST Task force report No. 32, Ames: CAST. http://www.cast-science.org/websiteUploads/publicationPDFs/acrylamide_ip.pdf; 2006

[95] Mestdagh F, De Wilde T, Fraselle S, Govaert Y, Ooghe W, Degroodt JM., Verhe R, Van Peteghem C, De Maulenaer B. Optimization of the blanching process to reduce acrylamide in fried potatoes. LWT-Food Science and Technology. 2008;**41**:1648-1654. DOI: 10.1016/j.lwt.2007.10.007

[96] Mestdagh F, Maertens J, De Wilde T, Cucu T, Delporte K, Van Peteghem C, De Maulenaer B. Chemical pre-treatments of potato products: Mechanisms of acrylamide mitigation and effects on the sensory quality. Communications in Agricultural and Applied Biological Science. 2007;**72**:9-12

[97] Timilsena YP, Khanal JS, Anal AK. Acrylamide: Thermally induced toxicant in foods and its control measures. Journal of Food Science and Technology Nepal. 2010;**6**:19-30. DOI: 10.3126/jfstn.v6i0.8256

[98] FDE (Food Drink Europe). Acrylamide Toolbox 2013. http://www.fooddrinkeurope.eu/uploads/publications_documents/AcrylamideToolbox_2013.pdf; 2013

[99] Haase NU. The formation of acrylamide in potato products. In: Skog K, Alexander J, editors. Acrylamide and Other Hazardous Compounds in Heat-Treated Foods. Cambridge, England: Woodhead Publishing Limited; 2006. pp. 41-56

[100] FDA (Food and Drug Administration). Numerical Listing of GRAS Notices. US Food and Drug Administration, Center for Food Safety and Applied Nutrition (CFSAN). http://www.foodsafety.gov/~rdb/opa-gras.html; 2009. Maryland, USA

[101] Fredriksson H, Tallving J, Rosen J, Aman P. Fermentation reduces free asparagine in dough and acrylamide content in bread. Cereal Chemistry. 2004;**81**:650-653. DOI: 10.1094/CCHEM.2004.81.5.650

[102] ARSO (African Organization for Standardization). Reduction of Acrylamide in Potato Products; 2012. Nairobi, Kenya. Reference No. ARS 849:2012(E), ICS 67.080.01

[103] Foot RJ, Haase NU, Grob K, Gondé P. Acrylamide in fried and roasted potato products: A review on progress in mitigation. Food Additives and Contaminants. 2007;**24**:37-46. DOI: 10.1080/02652030701439543

[104] Abong GO, Okoth MW, Karuri EG, Kabira JN, Mathooko FM. Levels of reducing sugars in eight Kenyan potato cultivars as influenced by stage of maturity and storage conditions. Journal of Animal and Plant Science. 2009;**2**:76-84

[105] Abong GO, Kabira JN. Potential food safety concerns in fried potato products in Kenya. Open Access Library Journal. 2015;**2**:1-11. DOI: 10.4236/oalib.1101522

[106] Sowokinos JR. Allele and isozyme pattern of UDP-glucose pyrophosphorylase as a marker for cold - sweetening resistance in potatoes. American Journal of Potato Research. 2001:**78**: 57-64. DOI: 10.1007/BF02874825

[107] Kahkeshani N, Saeidnia S, Abdollahi M. Role of antioxidants and phytochemicals on acrylamide mitigation from food and reducing its toxicity. Journal of Food Science and Technology. 2015;**52**:3169-3186. DOI: 10.1007/s13197-014-1558-5

[108] Zhang Y, Zhang Y. Formation and reduction of acrylamide in Maillard reaction: A review based on the current state of knowledge. Critical Reviews in Food Science and Nutrition. 2007;**47**:521-542. DOI: 10.1080/10408390600920070

[109] Hamzalioglu A, Mogol BA, Lumaga RB, Fogliano V, Gokmen V. Role of curcumin in the conversion of asparagine into acrylamide during heating. Amino Acids. 2013;**44**: 1419-1426. DOI: 10.1007/s00726-011-1179-5

[110] Jin C, Wu X, Zhang Y. Relationship between antioxidants and acrylamide formation: A review. Food Research International. 2013;**51**:611-620. DOI: 10.1016/j.foodres.2012.12.047

[111] Zhang Y, Chen J, Zhang X, Wu X, Zhang Y. Addition of antioxidant of bamboo leaves (AOB) effectively reduces acrylamide formation in potato crisps and French fries. Journal of Agricultural and Food Chemistry. 2007;**55**:523-528. DOI: 10.1021/jf062568i

[112] Fernandez S, Kurppa L, Hyvonen L. Content of acrylamide decreased in potato chips with addition of a proprietary flovoniod spice mix (Flavomare®) in frying. Innovations in Food Technology. 2003;**2**:24-26

[113] Key S, Ma JKC, Drake PMW. Genetically modified plants and human health. Journal of the Royal Society of Medicine. 2008;**101**:290-298. DOI: 10.1258/jrsm.2008.070372

[114] Curtis TY, Powers SJ, Balagiannis D, Elmore JS, Mottram DS, Parry MAJ, Raksegi M, Bedo Z, Shewry PR, Halford NG. Free amino acids and sugars in rye grain: Implications for acrylamide formation. Journal of Agricultural and Food Chemistry. 2010;**58**:1959-1969. DOI: 10.1021/jf903577b

[115] Chawla R, Shakya R, Rommens CM. Tuber-specific silencing of asparagine synthetase-1 reduces the acrylamide-forming potential of potatoes grown in the field without affecting tuber shape and yield. Plant Biotechnology Journal. 2012;**10**:913-924. DOI: 10.1111/j.1467-7652.2012.00720.x

[116] Ye J, Shakya R, Shresth, P, Rommens CM. Tuber-specific silencing of the acid invertase gene substantially lowers the acrylamide-forming potential of potato. Journal of Agricultural and Food Chemistry. 2010;**58**:12162-12167. DOI: 10.1021/jf1032262

Chapter 8

The Use of Acrylic-Based Polymers in Environmental Remediation Studies

Aslı Erdem Yayayürük

Additional information is available at the end of the chapter

http://dx.doi.org/10.5772/intechopen.69880

Abstract

Heavy metals are not biodegradable and tend to accumulate in living organisms. Many heavy metal ions are known to be toxic or carcinogenic. Thus, removal of these toxic heavy metals from water is of crucial importance to protect the human population and the environment. In recent years, polymeric adsorbents have received considerable interest for heavy metal removal mainly due to important technological and scientific developments such as easy synthesis at controlled dimensions with variable functional groups, perfect mechanical rigidity, tunable surface chemistry, large surface area, pore size distribution, high uptake values and feasible regeneration under mild conditions. This chapter has attempted to present to the readers the widespread investigations of acrylic-based polymeric adsorbents so that the reader can get an idea about the various types and forms of polymeric materials used for the removal of heavy metals from water.

Keywords: acrylics, heavy metal, removal, adsorption, remediation

1. Introduction

Remediation of environmental sources is a very important concern for human beings since these are essential to sustain life [1]. The presence of heavy metal ions in the environment is becoming a serious threat to public health and the environment. Heavy metal contamination exists in aqueous wastes of many industries, such as metal plating, mining operations, tanneries, chlor-alkali, radiator manufacturing, smelting, alloy industries and storage battery manufacture [2]. Most of the heavy metals are toxic, and therefore, their removal from environmental sources is very important [3]. Several methods can be used for the removal of metal ions from water including chemical precipitation, ion exchange, filtration, reverse osmosis, adsorption and membrane separation [4]. The advantages and the disadvantages of the various treatment processes used in heavy metal removal are given in **Table 1** [5].

By far, the most widely used method for removing heavy metals from solution is chemical precipitation. The soluble metal is converted into an insoluble precipitate (i.e. its hydroxide) by the increase in the pH of the solution. It is relatively simple and inexpensive to operate. The precipitates can be separated from the water by sedimentation or filtration. The treated water was then decanted and appropriately discharged or reused. The conventional chemical precipitation processes include hydroxide precipitation and sulphide precipitation [5].

Ion exchange is another most widely used method for the removal of heavy metals from the solution [6]. The ion exchanger is a solid capable of exchanging either cations or anions from the surrounding materials. Commonly used matrices for ion exchange are synthetic organic ion exchange resins.

Coagulation-flocculation can also be employed to treat water with heavy metals, wherein the coagulation process destabilises colloidal particles by adding a chemical agent (coagulant) and results in sedimentation [7]. Many coagulants are widely used in the conventional wastewater treatment processes such as aluminium, ferrous sulphate and ferric chloride, resulting in the effective removal of wastewater particulates and impurities by charge neutralisation of particles and by enmeshment of the impurities on the formed amorphous metal hydroxide precipitates. Coagulation is followed by flocculation of the unstable particles in order to increase their size and form into bulky floccules which can be settled out. Today, many kinds of flocculants, such as polyaluminium chloride, polyferric sulphate and polyacrylamide (PAM), are widely used in the treatment of waste water. However, it is nearly impracticable to remove heavy metals very well from waste water directly by these current flocculants.

Flotation is employed to separate solids or dispersed liquids from a liquid phase using bubble attachment [8]. Adsorptive bubble separation employs foaming to separate the metal impurities. Ion flotation, precipitate flotation and sorptive flotation are the main flotation process mechanisms for removal of metal ions from solution. Membrane filtration has received considerable attention for the treatment of inorganic effluent, since it is capable of removing not only suspended solid and organic compounds but also inorganic contaminants such as heavy metals.

Electrochemical methods involve the plating out of metal ions on a cathode surface and can recover metals in the elemental metal state. Electrochemical wastewater technologies involve relatively large capital investment and the expensive electricity supply. The cost factor prevents the usage of this method to be widely applied industrially [9].

Adsorption is now recognised as an effective and economic method for heavy metal wastewater treatment. The adsorption process offers flexibility in design and operation and in many cases will produce high-quality treated effluent. Among these methods, adsorption via the use of adsorbent is one of the most effective methods since it is rapid, economic, effective and easy. In addition, the regeneration of the adsorbent with resultant economy of operation may be possible because adsorption is sometimes reversible [10].

http://dx.doi.org/10.5772/intechopen.69880

Method	Advantages	Disadvantages
Chemical precipitation	Simple	Needs large amount of chemicals
	Inexpensive to operate	Excessive sludge production
	Convenient and safe	Sludge disposal cost
		High maintenance cost
		Slow metal precipitation
Ion exchange	High removal efficiency	Sensitive to the pH of the solution
	Fast kinetics	
	High treatment capacity	High initial cost
	High regeneration	High maintenance cost
Coagulation-flocculation	Bacterial inactivation capability	High chemical consumption
	Good sludge settling	High maintenance cost
	Good dewatering characteristics	High operation cost
		Increased sludge volume generation
Flotation	Removal of small particles	
	High selectivity to recover valuable ions	High maintenance cost
	Low retention time	High initial capital cost
	Production of more concentrated sludge	High operation cost
	High removal efficiency	
Membrane filtration	Low chemical consumption	Process complexity
	Low solid waste consumption	Limited use
	Small space requirement	High maintenance cost
	Possible to be metal selective	High initial capital cost
		High operation cost
Electrochemical treatment	Rapid	
	Less sludge production	High initial capital cost
	Less chemical use	High energy requirement
	Provide good reduction yields	Requires filtration for flocs
	Moderately metal selective	
Adsorption	High capacity	Performance depends on type of adsorbent
	Flexibility in design and operation	
	Fast kinetics	Chemical derivatization to improve sorption capacity
	Wide variety of target pollutant	

Table 1. Comparison of treatment processes used for heavy metal removal.

In recent years, polymeric adsorbents have received considerable interest for heavy metal removal mainly due to important properties such as easy synthesis at controlled dimensions with variable functional groups, perfect mechanical rigidity, tunable surface chemistry, large surface area, pore size distribution, high uptake values and feasible regeneration under mild conditions. These novel properties allow the polymeric adsorbent to be applied for use in drug delivery systems, optics and water treatment. These can also be incorporated with other particles making them extremely versatile [11].

2. Acrylic-based polymers

Acrylic monomers have been widely utilised for the preparation of polymeric materials that can be used as sorbents in environmental remediation studies due to their valuable characteristics such as water repellence, transparency and good film-forming ability. These have become more widely used in the last decades because they are also environmentally friendly, they offer easier clean-up, and, last but not least, their properties and application performance characteristics have been improved [12]. This section will explain the practical applications of polymeric adsorbents which are synthesised from their acrylic monomers for the removal of heavy metal ions in water.

2.1. Poly(acrylic acid) (PAAc)

Acrylic acid (AA), an important monomer of synthetic resins, is a versatile compound used in polymer synthesis. It has been widely used to fabricate a variety of functional materials, such as water-absorbent polymers, adhesives and textile-treatment agents and so on [13]. Radiation-induced graft polymerisation can be used to modify the polymers to introduce valuable properties to the synthesised materials. The technique is very advantageous over the conventional methods. For example, the chemical and photochemical grafting methods are simple and can be done without employing any additives or catalyst at any temperature [14].

Benamer et al. [15] have modified chitosan beads by a graft copolymerisation reaction of chitosan with AA by gamma radiation, and the removal of Cd and Pb ions from aqueous solutions was examined using the novel adsorbent. Raw chitosan beads were first modified by cross-linking with glutaraldehyde, and the radiation-induced grafting of AA onto the chitosan beads was performed using ^{60}Co gamma-source. The grafting rate of 81% was obtained at a dose rate of 20.64 Gy min^{-1}. The FTIR was used for the structural characterisation of the sorbent, and their swelling measurements were performed at different pH values. The adsorption experiments were carried out both with modified and unmodified chitosan beads for Cd(II) and Pb(II) ions, and it was found that the grafting of new groups increased the number of adsorption sites in chitosan, leading to the increase in adsorption capacity. The optimum pH for Cd(II) and Pb(II) uptake was between 4 and 7 using the synthesised adsorbent. The experimental kinetic data were found to have a better fit with pseudo-second-order reaction, and Langmuir isotherm model was found to be more suitable to represent the experimental equilibrium.

González-Gómez et al. [16] have synthesised two new comb-type hydrogels using two monomers such as AA and 4-vinyl pyridine (4VP) by gamma radiation. The synthesised adsorbents were as follows: (a) Comb-type hydrogels of an AA network followed by grafting of 4VP (net-PAA)-g-4VP and comb-type hydrogels of an AA network grafted onto polypropylene (PP) followed by grafting of 4VP [net-(PP-g-AA)-g-4VP]. The ^{60}Co gamma source was used to synthesise the comb-type systems. The AA in synthesised comb-type hydrogels was capable of removing metals because the ions primarily interact with the carboxyl groups. Moreover, the amino groups in 4VP increased the stability constant and complexation bond strength between the metals and functional groups. The retention of Cu(II) and Zn(II) ions was evaluated by the new adsorbent, and a high adsorption capacity was found through the target ions. The quantification of the metal ions was performed with a colorimetric method using ultraviolet visible (UV-Vis) spectrophotometry. The net-(PP-g-AA)-g-4VP system exhibited better efficiency with Zn(II) ions (up to 90%), whereas the (net-PAA)-g-4VP system exhibited good efficiency for retaining both ions, with approximately 85–90% retention. The equilibrium adsorption capacities of Zn(II) ion were found to be 1086 and 480 mg g^{-1} for net-(PP-g-AA)-g-4VP and (net-PAA)-g-4VP, respectively, which were higher than some results reported in literature. Moreover, adsorption/desorption studies indicated that the regeneration of the adsorbent was possible by altering the pH of the solutions. Freundlich model was the best model to describe the adsorption process, indicating heterogeneous adsorption. In addition, the kinetic results agreed with pseudo-second-order reaction indicating that the retention process occurred by chemisorption.

Radiation-induced graft copolymerisation of AA, styrene (Sty) on poly(N-vinyl-2-pyrrolidon) (PVP), was realised by El-Mohdy et al. [17] to remove Co(II), Cu(II) and Fe(III) metal ions from aqueous solutions. In order to prepare the hydrogels, AA and Sty were mixed together with different monomer feed ratios (AAc:Sty) at room temperature. The PVP (50 wt%) was mixed with these compositions and then irradiated with ^{60}Co gamma rays at a dose rate of 6.13 kGy/h. The obtained hydrogel was washed with ethanol/water mixture to remove unreacted monomer components and then dried in vacuum oven at 40°C. The sulfonation (Sf) (to introduce sulfonic group onto the styrene benzene ring) and alkaline treatment (to convert –COOH group of AA into –COO) were applied in order to functionalise the hydrogels. The success of characterisation of the newly synthesised materials was confirmed using FTIR, UV, thermogravimetric analysis (TGA), scanning electron microscopy (SEM) and X-ray diffraction (XRD) methods. The adsorption experiments of Co(II), Cu(II) and Fe(III) ions from aqueous solutions using the hydrogels were studied in terms of pH, temperature, initial metal concentration, contact time and polymer structure. The metal ion uptake efficiency of PVP/(AA-co-Sty) hydrogel was found to be in the following order: hydrogel treated with NaOH/Sf > hydrogel treated with NaOH > hydrogel treated with Sf > untreated hydrogel. The highest metal ion uptake was observed for Fe(III), and the lowest was observed for Cu(II). It was concluded that the hydrogels could be used as a potential adsorbent for the removal of target ions in environmental studies.

The poly(acrylic acid) (PAA), a polymer containing polar carboxylic functional groups in linear CH_2-CH_2 chain, has a great ability to form complexes with metal ions. It is pH sensitive and belongs to the class of commercial polymers produced on a large scale. It is widely used in various industries, such as agriculture, medicine, etc., as pharmaceutical carriers, antifouling agents, flocculants and super absorbents [18].

Sezgin and Balkaya [19] have investigated the adsorption of Cu(II), Ni(II), Zn(II) and total Cr [Cr(III) and Cr(VI)] ions using polyacrylic acid hydrogel. The adsorbent was synthesised by free radical participation polymerisation of acrylic acid monomer in the presence of 1 mol% monomer cross-linker [N,N-methylenebisacrylamide (NMBA)], 1 mol% monomer initiator [potassium persulphate ($K_2S_2O_8$)] and an equal weight of accelerator [potassium bisulphite ($KHSO_3$)] under nitrogen atmosphere. The polymeric hydrogel was characterised by FTIR. The surface characteristics of the hydrogel were evaluated using a scanning electron microscopy/energy dispersive X-Ray (SEM/EDX) analysis. The effects of adsorbent amount, contact time and shaking speed were studied using a real wastewater sample. Among the kinetic models investigated, pseudo-second-order model was found to describe the adsorption process since the correlation coefficients of the model are high, and the calculated q_e (amount of adsorbate adsorbed per unit mass of adsorbent at equilibrium) was in good agreement with the experimental ones. Moreover, a multiple diffusion model was determined to be effective in adsorption. Freundlich model was found to represent the adsorption data. Thermodynamic experiments have pointed out a spontaneous and exothermic behaviour. A comparison for the removal of the metal ions was performed for industrial wastewater (IWW) and synthetic wastewater (SWW) samples, and a higher adsorption capacity was found for SWW. According to the results presented in the manuscript, AA hydrogel was found to be as an effective material in order to remove target metal ions from wastewater samples.

The PAA hydrogel beads were also examined as a potential adsorbent for the removal of Pb(II)ion from waste water [20]. The batch-type adsorption was applied, and the effect of pH contact time and temperature was studied. The optimum pH value was found to be 6.3, and the adsorption of Pb(II) increased with increase in temperature and time. The maximum capacity was reached in 24 hours with 113.43 mg g^{-1} value. The adsorption kinetic data were described by the pseudo-first-order kinetic model, and the adsorption isotherm agreed well with the Langmuir model. Moreover, the adsorption process was stated to be spontaneous and endothermic.

Although PAA has a strong ability to form complexes with metal ions, the high water solubility of this polymer hinders its application to remove the metal ions from the aqueous solutions. To overcome these limitations, the polymer was modified with several functional groups or combined with various materials such as clay, chitosan, carbon nanotubes, etc. [21, 22].

A polyacrylic acid–organobentonite nanocomposite (PAA-Bent) was synthesised by the intercalation of cetyl trimethylammonium (CTA) surfactant and polyacrylic acid into the bentonite interlayer spaces. The XRD patterns and FTIR evidences of the novel adsorbent confirmed the success of the synthesis [23]. The newly synthesised material was applied for the removal of Pb(II) ions, and a removal efficiency of 99.6% was achieved by the nanocomposite with a 30-minute contact time for 7.5 g L^{-1} solid-to-liquid ratio and an initial metal concentration of 400 mg L^{-1} at 25°C. The maximum adsorption capacity of the adsorbent was 93 mg g^{-1} which was approximately twice as much as that of untreated Bent (52 mg g^{-1}). It was stated that the superior capacity and affinity of PAA-Bent nanocomposite towards Pb(II) were probably due to the complex formation between the polymer carboxyl groups and Pb(II) ions. The kinetics and thermodynamics of Pb(II) adsorption onto bentonite and PAAc-Bent adsorbent were also investigated by the same research group in another study [24]. Batch-type adsorption experiments were conducted at different times, Pb(II) concentrations and temperatures. A fast adsorption (≥99.4% adsorption in 30 min) was observed at low lead concentration due to high complexation affinity between metal ions and reactive functional groups on the surface of PAA-Bent. At higher lead concentrations, more Pb(II) ions were left unabsorbed in solution due to the saturation of binding sites so that the equilibrium time for Pb(II) adsorption by the PAA-Bent nanocomposite increased from ~30 to ~60 min. When the effect of temperature was examined, the adsorption of lead ions increased with increasing temperature both for Bent and PAA-Bent. The equilibrium studies showed that Pb(II) adsorption data on PAA-Bent followed the Langmuir model. The endothermic and spontaneous adsorption of Pb(II) ions for both Bent and PAA-Bent adsorbents was evaluated using thermodynamic studies. The adsorption kinetics of Pb onto PAA-Bent fitted better with the pseudo-second-order and pseudo-first-order kinetic models better than the Elovich and intraparticle diffusion models suggesting that chemi-adsorption controls the adsorption process. These results have indicated that PAA-Bent nanocomposite could be efficiently used as an adsorbent in lead removal studies from waste water.

Shirsath et al. [25] have prepared PAA hydrogel composite by incorporating kaoline (K) clay using ultrasound-induced polymerisation and conventional process. The utility of the synthesised hydrogels has been investigated for the removal of brilliant green (BG) dye. The PAA-K hydrogel was synthesised using AA (monomer), ammonium persulphate (initiator) and sodium dodecyl sulphate (surfactant). The morphology of PAA-K hydrogel was evaluated using FTIR and transmission electron microscope (TEM). The particle size of kaolin clay was measured from the TEM images and found to be in the range of 20–50 nm. The hydrogels were prepared using ultrasound-induced polymerisation and conventional process. The effect of different parameters (pH, temperature, initial dye concentration, quantity of hydrogel and clay content) on the extent of adsorption was examined in batch mode for the hydrogels that were prepared both by using ultrasound-induced polymerisation and conventional process. The concentration of BG dye was measured using UV-Vis spectrophotometer at 624 nm. Better results were obtained for the hydrogel prepared by ultrasound-assisted polymerisation process. The removal of BG dye was found to increase with an increase in pH (till the optimum value of 7), temperature, adsorbent and adsorbate concentrations as well as the kaolin clay content in the hydrogel. It has been observed that the maximum BG dye removal by PAA-K hydrogel was achieved for the initial dye concentration of 30 mg L^{-1} at a temperature of 35°C and pH 7. The pseudo-second-order kinetic model was found to fit the kinetic data. Both Freundlich and Langmuir adsorption isotherm models explained the experimental results satisfactorily. Moreover, the adsorption was spontaneous and endothermic in nature.

The removal of Co(II) from the aqueous solutions was investigated using PAA-grafted multiwall carbon nanotubes (MWCNTs) by Chen et al. [26]. Plasma techniques were applied to graft PAA onto the surface of MWCNTs to improve their dispersion property and adsorption capacity in the aqueous solutions. The characterisation of both raw MWCNT and MWCNT-g-PAA was performed using techniques such as FTIR, TGA, Raman spectroscopy, SEM and TEM. The microstructures of the raw MWCNTs and MWCNT-g-PAAc were observed by SEM and TEM. An entangled and curled framework was noticed for both of the adsorbents with almost no differences in their surface. A more compact stacking morphology was observed for MWCNT-g-PAA due to the interactions of carboxylic acid groups on the surface of MWCNTs. A very thin layer of grafted PAA was coated on partial external walls of MWCNTs and was investigated with TEM images. The results of FTIR spectra of raw MWCNTs and MWCNT-g-PAA were compared, and MWCNT-g-PAA exhibited a relatively stronger broad peak around 3400 cm^{-1} due to the O–H stretching vibration, which suggests the existence of hydroxyl, the functional group from PAA and chemisorbed water. Moreover, peaks at 3430 cm^{-1} (—OH), 1680 cm^{-1} (C=O), 700 cm^{-1} ($-CH_2-$) and 1080 and 1200 cm^{-1} (C—O) suggest that PAA was successfully grafted on the surface of raw MWCNTs by the plasma-induced grafting method. The TGA curves have indicated that the carbon impurity (such as amorphous carbon) in MWCNTs used was negligible, and the weight percentage of grafted PAA on the surface of MWCNTs was estimated to be 4.4%. The adsorption capacity of both adsorbents towards Co(II) ions was investigated, and the adsorption parameters (shaking time, pH, ionic strength and temperature) were optimised. A pH-dependent and rapid adsorption was observed. The main adsorption mechanism for Co(II) adsorbed onto MWCNT-g-PAA was stated to be due to the complexation of Co(II) with carboxyl groups on the surface of MWCNT-g-PAA. The spontaneous and endothermic adsorption process was well described by the pseudo-second-order model. As a result, PAA was successfully grafted on the surface of raw MWCNTs and obviously enhanced its adsorption capacity for Co(II) which enables MWCNT-g-PAA as a promising adsorbent in water purification studies.

The effect of contact time, pH, foreign ions and PAA on oxidised MWCNT for the adsorption of Ni(II) was studied by Yang et al. [27]. The adsorbent characterisation was confirmed using FTIR. A rapid equilibration was achieved within 2 hours, and the experimental data were found to fit well by the pseudo-second-order model. The adsorption of Ni(II) on oxidised MWCNTs increased with increasing pH. A positive effect of PAA on Ni(II) adsorption was found at $pH < 8$, whereas a negative effect was observed at $pH > 8$. The effect of addition sequences of PAA/Ni(II) on the adsorption of Ni(II) to PAA-MWCNT hybrids was also studied. The results indicated that the adsorption of Ni(II) was influenced by the addition sequences obviously. The Ni(II) adsorption on oxidised MWCNTs was dependent on foreign ions at low pH values and independent of foreign ions at high pH values. The adsorption of Ni(II) on oxidised MWCNTs was stated to complicate and was attributed to surface complexation and ion exchange. The synthesised materials were found to be suitable adsorbents for the solidification and preconcentration of Ni(II) from aqueous solutions.

2.2. Polyacrylamide (PAM)

Polyacrylamide (PAM) is a water-soluble polymer, containing large number of amide groups as side groups, can be grafted onto various matrices and be used in environmental remediation studies as adsorbent [28].

Liu and Guo [29] have synthesised PAM-grafted attapulgite (ATP) nanofibrils by the surface-initiated atom transfer radical polymerisation of acrylamide from the modified surfaces of the fibrillar clays. The elemental analysis, FTIR, TGA and TEM were used to characterise the synthesised adsorbent. The percentage of grafting (PG%) was calculated from the results of the elemental analysis, and polymerisation time has been found to increase PG%. The adsorption capacity of both bare ATP and PAM-ATP was evaluated both for mercury and dyes (methylene blue and methyl orange). It was evaluated that surface graft polymerisation of acrylamide enhanced the adsorption capacity towards mercury and methylene blue but not methyl orange.

Kaşgöz et al. [30] have modified PAM hydrogels by Mannich reaction using different amine compounds, namely, diethylenetriamine (DETA), ethylenediamine (EDA) and triethylenetetramine (TETA), and sulfomethylation reaction. The FTIR was used to characterise the adsorbents. Moreover, basic group content (BGC) and hydroxymethyl group content (HMG) and the equilibrium degree of swelling (EDS) were evaluated. These results indicated that the amine value (AV) of the adsorbents increased with time, temperature and amine amount. The adsorbents synthesised were used for the removal of Cu(II), Cd(II) and Pb(II) ions in competitive or non-competitive conditions, and the effect of pH and adsorption rate was investigated. It was observed that the removal capacities of the adsorbents changed mainly according to their AV, BGC and EDS. Moreover, the Mannich products were selective towards Cu(II) ion, and the sulfomethylation products were highly selective to Pb(II) ions. Furthermore, regeneration studies were performed to demonstrate the practical use of the adsorbents.

The hydroxamic acid–modified polyacrylamide/Fe_3O_4 adsorbent (M-PAM-HA) was synthesised by Zhao et al. by microemulsion polymerisation with acrylamide determined [31]. Its structure was characterised by various methods (XRD, FTIR, TGA, SEM, EDS, zeta potential), and its swelling and iron-leaching properties were also determined. Batch-type adsorption experiments were applied in order to investigate the removal efficiency of the adsorbent towards Cd(II), Pb(II), Co(II) and Ni(II) ions. The adsorption kinetic experiments revealed that the results are best correlated with the pseudo-second-order model indicating that the rate-controlling step was chemisorption. Moreover, the equilibrium data have shown that Sips isotherm model best fitted with the experimental results rather than Freundlich and Langmuir isotherm models. The adsorption mechanism was elucidated both with FTIR and theoretical calculations, and it was proposed that the carbonyl and hydroxyl groups formed stable five-membered ring chelates with divalent metal ions. Moreover, 0.1 M Na_2EDTA was found to regenerate the adsorbent efficiently which shows that the novel adsorbent has great potential in wastewater treatments.

In a study by Yavuz et al. [32], polyacrylamide was grafted onto cross-linked poly(4-vinyl pyridine) (P4-VP-g-PAm) and used only for the purpose of mercury removal from water samples at mg L^{-1} concentration levels. Yayayürük et al. [33] have used the same adsorbent, for the first time, for the selective preconcentration of Hg(II) prior to cold-vapour atomic-fluorescence measurements, and separate determinations of Hg(II) and MeHg(I) at the trace level were performed at a ng L^{-1} level. In addition, adsorption isotherms were studied, and the affinity of the adsorbent to several metal ions, such as Pb(II), Zn(II), Cu(II), Cd(II) and Fe(III), was investigated. The developed method was also applied to the determination of Hg(II) and MeHg(I) in sea water and estuarine water. The high uptake capacity of the adsorbent allowed the quantitative retention of Hg(II) in water samples, which is very important in Hg-polluted matrices.

2.3. Poly(glycidyl methacrylate) (PGMA)

The use of poly(glycidyl methacrylate) (GMA) has become very attractive as an adsorbent in environmental remediation studies during the last years because of its high tensile strength, porous structure and resistance to acid and alkaline media. Moreover, GMA can be readily modified by chemical reaction and grafting. The GMA functional group can be easily transformed into urea, pyrazole, amines, iminodiacetic, pyridine groups, etc. due to the epoxy group in GMA molecule [34].

Nastasovic et al. [35] have synthesised macroporous cross-linked poly(glycidyl methacrylate-co-ethylene glycol dimethacrylate) and poly(GMA-co-EGDMA) with different porosities by suspension copolymerisation and functionalised with ethylene diamine and diethylene triamine. These have elucidated the kinetics of the adsorbents towards hexavalent chromium [Cr(VI)] under both competitive and non-competitive conditions. The competitive kinetics was studied with several multicomponent solutions. A very fast Cr(VI) uptake was observed in non-competitive conditions due to the amino groups present at the adsorbent's surface. It was found that the adsorption of Cr(VI) was much slower from their binary solutions than from single-component solution. This situation was pronounced due to competition of the metals for the active sites of the adsorbents. The pseudo-second-order model was determined to best fit with adsorption data meaning that adsorption depends both on the properties of the metal and the adsorbent. Nastasovic et al. [36] have also functionalised macroporous poly(glycidyl methacrylate-co-ethylene glycol dimethacrylate) (PGME) by reacting the pendant epoxy groups with diethylene triamine (PGME-10/12-DETA) and investigated the adsorption mechanism of the adsorbent towards Cu(II), Cd(II) and Pb(II) ions in aqueous solution. The FTIR, X-ray photoelectron spectroscopy (XPS), SEM-EDX and atomic force microscopy (AFM) were used to elucidate the adsorption mechanism before and after interaction of PGME-10/12-DETA with the target ions. The XPS and FTIR analyses suggest that complexation through the formation of metal–O and metal–N bonds with the -OH, -NH and $-NH_2$ groups as the possible mechanism of Cu(II), Cd(II) and Pb(II) adsorption on sample PGME-10/12-DETA. The AFM images indicate that the metal adsorption induces change in the size and morphology of the amino-functionalised PGME. The time required to reach 50% of the total adsorption capacity was 5 min for all the investigated metal ions. Already after 30 min, PGME-10/12-DETA attained approximately 90%, and after 180 min, 95% of metal ions with maximal capacity was adsorbed. The pseudo-first-order, pseudo-second-order and surface reaction-based kinetic models were fitted to the experimental data, and the pseudo-second-order model was found to describe adsorption kinetics for all metal ions.

A magnetic glycidyl methacrylate resin was synthesised by Elwakeel et al. [37] for the removal of Hg(II) in industrial waste water. The adsorbent was synthesised by coating a magnetite core with glycidyl methacrylate–based polymer. Further, it was grafted with diethylene triamine (DETA). The characterisation methods (SEM, SEM-EDX, FTIR, BET and TEM) used indicated the successful synthesis of the polymer. The absence of coercivity and remanence in vibrating sample magnetometry (VSM) analysis confirmed that the material has super paramagnetic properties. The effects of pH, counter ions, adsorbent amount and the temperature were investigated. Moreover, reaction kinetics and adsorption isotherms were studied.

The maximum adsorption capacity was obtained at a pH of ~4. The adsorption kinetics and isotherms were best fitted with the pseudo-second-order model and Langmuir equation, respectively. Thermodynamic studies indicated that adsorption of Hg(II) by the adsorbent is exothermic and spontaneous with increased randomness. The desorption from the adsorbent was realised with KI or KI-thiourea mixture, and at least four cycles of adsorption/desorption were obtained with the proposed adsorbent. Two industrial effluents were used for the removal of Hg(II), and satisfactory results were obtained.

In a recent work by Şenkal and Bıçak, glycidyl methacrylate (GMA)-methyl methacrylate (MMA)-divinylbenzene (DVB) terpolymer having diethylenetriamine tetra acetic acid (DTTA) as a functional group was synthesised and used for the efficient removal of Ca(II) and Mg(II) ions from water [38]. The cross-linked terpolymer beads were prepared by suspension polymerisation of GMA (0.4 mol), MMA (0.5 mol) and DVB (0.1 mol) mixture and have been modified through epoxy functional groups in two steps: (i) by treating with excess of diethylene triamine (DETA) and (ii) by subsequent reaction with potassium chloroacetate. The resulting polymer possesses DTTA functional groups (with a degree of functionalisation, DF: 1.70 mmol g^{-1}). The polymeric adsorbent was separated into different particular sizes within the range of 125–250 μm by steel sieves prior to use. This novel adsorbent was also used for the removal of Hg(II) and Cu(II) ions in aqueous solution. In order to clarify the adsorption process with the novel adsorbent, the effect of pH, contact time and adsorbent amount was elucidated. The adsorption process was then investigated in terms of Langmuir and Freundlich isotherms to characterise the uptake of target ions by the adsorbent. In addition, the affinity of the adsorbent to several metal ions was investigated. The developed method was also applied to the determination of Hg(II) and Cu(II) in water samples. According to the obtained results, the polymeric adsorbent was proposed as a promising adsorbent for the removal of target ions in contaminated water [39, 40].

2.4. Poly(methyl methacrylate)

Poly(methyl methacrylate) (PMMA) is an important member in the family of poly(acrylic ester(s)). It was found to have widespread use due to its desirable properties such as good solvent resistance, excellent optical transparency, low cost and good flexibility [41].

Al-Muhtaseb et al. [42] have evaluated the adsorption potential of PMMA to remove phenol from water. The PMMA was polymerised using atom transfer radical polymerisation (ATRP) technique. The percentage conversion of the methyl methacrylate monomer was determined by weighing the dried polymer. Both ^{1}H and ^{13}C nuclear magnetic resonance (NMR) spectra of the adsorbent were recorded. Moreover, FTIR spectra of the adsorbent were taken before and after adsorption with phenol to check the functional groups of the materials. Batch-type adsorption was applied at various temperatures (25–55°C), initial phenol concentrations (10–90 mg L^{-1}) and contact times (15 min to 4.5 hours). Phenol adsorption was found to increase with increase in contact time and initial phenol concentration. The data were best correlated with the pseudo-second-order kinetic model. From the intraparticle diffusion and Boyd models, the adsorption of phenol on PMMA was shown to be governed by film diffusion. Both Freundlich and Redlich-Peterson isotherm models were found to describe the adsorption data. Thermodynamic parameters indicated that the adsorption process is exothermic and spontaneous in nature and higher ambient temperature results in more favourable adsorption [42].

Although PMMA is non-toxic, easy to obtain, cheap and being insoluble in water, its brittle texture and low thermal and mechanical stability are the disadvantages of this material. To overcome these drawbacks, several chemicals such as mesoporous silica or carbon were added to the polymer matrix that will result in an increase both in thermal and mechanical behaviour and its physical performance. Salisu et al. [43] have synthesised a novel adsorbent, sodium alginate-graft-poly(methyl methacrylate) (Alg-g-PMMA) in the form of calcium cross-linked beads and utilised for the removal of Pb(II) in aqueous solution. The beads were formed using a simple encapsulation method that enables the ionic interactions between glutaronate blocks of the graft copolymer and calcium ions. The FTIR, SEM and particle size analyser were used to characterise the adsorbents. The beads owing a spherical shape, rough surface and porosity having a mean size of ~1.2 mm were obtained. The adsorption process was found to depend on the initial pH and Pb(II) concentration. The optimum pH for the adsorption of the metal ions was found to be 4.0. The Langmuir model was found to describe the adsorption data with a maximum adsorption capacity of 526 mg g^{-1}. Moreover, the adsorption followed a pseudo-second-order kinetic model. HNO_3 (0.1 M) was used in the regeneration studies. The results showed that the material proved to be excellent in the removal of the Pb(II) ions after nine cycles with no significant loss of adsorption capacity.

The oxidative-free radical-graft copolymerisation reaction was applied in order to synthesise a novel adsorbent, poly(methyl methacrylate)-grafted alginate/Fe_3O_4 nanocomposite for the removal of Pb(II) ions in aqueous solution [44]. The FTIR, XRD, SEM, TEM, TGA and DSC were employed to characterise the adsorbent. The adsorption feasibility of the adsorbent towards Pb(II) and Cu(II) ions was investigated in terms of pH, contact time, initial metal concentration and adsorbent dose. The adsorption of Pb(II) and Cu(II) was found to be pH dependent, and a pH 5 was found to be optimum for both of the metal ions. An increase was observed in the adsorption capacity of the adsorbent as the metal concentrations increase. This was due to the electrostatic interactions between metal ions and adsorbent, which involves active sites of progressively lower affinity for the heavy metal ions up to the saturation point. Moreover, the amount of metal ions was adsorbed rapidly at the initial stages of the adsorption process due to the availability of binding sites of the adsorbent, but as time went by, the adsorption slowed down before reaching equilibrium at about 180 min. The adsorption capacity of the adsorbent was increased with increasing the temperature indicating an endothermic nature of the adsorbent. The pseudo-first-order and pseudo-second-order Elovich and Weber-Morris intraparticle diffusion models were investigated to elucidate the adsorption kinetics, and the data were best described with the pseudo-second-order kinetic model. The experimental data were also investigated using Langmuir, Freundlich, Sips and Temkin models, and the data were best followed by the Freundlich model. The desorption of Pb(II) and Cu(II) was carried out by using 0.1 M HCl solution. The comparison of the adsorbent was done with some studies in literature, and it was found that the adsorbent exhibited a very good adsorption capacity towards Pb(II) and Cu(II). Therefore, the present adsorbent can be utilised for successful removal of these metal ions from industrial wastewater samples.

The poly(ethylene imine) (PEI)-immobilised PMMA microspheres were synthesised by Duru et al. [45] by suspension polymerisation and subsequent modifications. The SEM and elemental analysis were performed to characterise the adsorbent. The free amine content of the PEI-immobilised PMMA microspheres was determined by potentiometric titration. The extent of removal of copper, cadmium and lead ions was evaluated with different initial amounts of metal ions (50–600 mg L^{-1}) and at several pH values (3.0–7.0). The adsorption of heavy metal ions on the unmodified PMMA microspheres was very low. But, PEI-immobilised PMMA microspheres exhibited a higher adsorption capacity due to the presence of the metal-chelating ligand PEI on the surface of the adsorbent. The adsorption capacities increased with increasing pH, and an optimum pH value of 5.5 for Cu(II) and Pb(II) and 6.0 for Cd(II) was evaluated. The affinity order of adsorption was Cd(II) > Cu(II) > Pb(II). The experimental data were found to be consistent with the Langmuir model.

2.5. Poly(n-butyl acrylate) (PBA)

Poly(n-butyl acrylate) can also be used in adsorption studies since its homopolymer had a low glass transition temperature and hydrophobic characteristics. These two important properties make it to be tailored to several materials (e.g. chitosan and zeolite), and thereby, the mechanical properties and water resistance properties will be improved in the final product [46].

Kumar et al. [47] have applied a novel microwave-assisted method for the preparation of n-butyl acrylate–grafted chitosan and used it as an adsorbent in Cr(VI) removal studies. Three minutes of irradiation time was applied to prepare the adsorbent. The analysis of chromium was performed spectrophotometrically at 540 nm by complexation of Cr(VI) with diphenylcarbazide. The surface characterisations of the adsorbent and the mechanism of the adsorption process were achieved using techniques, namely, FTIR, XRD and SEM/EDS. Moreover, XPS spectra of the n-butyl acrylate-grafted chitosan before and after adsorption with Cr(VI) were assessed. It was evaluated that the adsorption mechanism involved the electrostatic interaction between hydrochromate anion and the amino hydroxyl groups of the adsorbent. An effective adsorption was achieved at a pH of 3.5 with 25 mL of 20 mg L^{-1} Cr(VI) solution. The Langmuir, Freundlich, Dubinin-Radushkevich, Temkin, Elovich and Redlich isotherms were studied in detail. A Langmuir adsorption capacity of 17.15 mg g^{-1} was attained, and the negative entropy change obtained in the study was due to the decrease in randomness at adsorbent-solution interface. The pseudo-second-order kinetics and spontaneous exothermic adsorption process were evaluated in the study. The quantitative desorption of chromium was achieved with ammonium hydroxide. Finally, n-butyl acrylate-grafted chitosan was proposed as a promising adsorbent in chromium removal in real samples.

The Cr(VI) removal was also studied using chitosan-g-poly(butyl acrylate)/silica gel (Cs-g-PBA/SG) nanocomposite in aqueous solution. The modification of chitosan was achieved by graft copolymerisation with butyl acrylate, and glutaraldehyde cross-linked silica gel/chitosan-g-poly(butyl acrylate) nanocomposite was synthesised by sol-gel method. The particle size of the prepared Cs-g-PBA/SG nanocomposite was found to be 615.1 nm. The BET-specific surface area and average pore radius were determined by BET experiment and found to be 95.9044 m^2 g^{-1} and 1.7787 nm, respectively. The FTIR spectrum of bare Cs-g-PBA copolymer and Cs-g-PBA/SG nanocomposite before and after chromium adsorption was studied to analyse the reaction routes of the modification and adsorption process. The surface morphology of the Cs-g-PBA/SG nanocomposites obtained with SEM indicated that the nanocomposite was composed of particles of different shapes and sizes. The surface characterisation methods appeared to indicate that chromium ions were adsorbed via binding to the abundant –OH, –NH and C=O sites available in the composite prepared using sol-gel process which offered the possibility to tailor silica porosity and modulate the diffusion properties of the composites. Batch-type experiments were also conducted to investigate the effect of contact tine, adsorbent dose and pH, and the optimum values were found to be 120 min, 4 g and 7.0, respectively. The Langmuir model agreed well with the experimental data with a maximum adsorption capacity of 55.71 mg g^{-1}. Moreover, the kinetic data followed pseudo-second-order model. The authors have concluded that Cs-g-PBA/SG nanocomposite is an excellent adsorbent for Cr(VI) removal from waste water [48].

Nithya and Sudha have also synthesised chitosan-g-poly(butyl acrylate)/bentonite nanocomposite and used for the removal of Cr(VI) and Pb(II) ions from tannery industrial waste water. The samples collected were also analysed for total hardness, turbidity, total solids, total dissolved solids, total suspended solids, chemical oxygen demand, biochemical oxygen demand, salinity and electrical conductivity. Batch-type adsorption experiments were conducted, and the influence of different experimental parameters on adsorption, such as pH (2–8), contact time (1–6 hours) and adsorbent amount (1–6 g), was evaluated. The maximum removal percentage (97.81%) of Cr(VI) in the tannery industrial waste water was found at an optimum adsorbent dosage of 5 g, contact time of 4 hours and pH of 3. Therefore, chitosan-g-poly(butyl acrylate)/bentonite nanocomposite was proposed as a promising adsorbent for target ions removal [49].

2.6. Poly(2-hydroxyethyl methacrylate) (PHEMA)

2-Hydroxyethyl methacrylate (HEMA) is an important functional monomer which is widely used in the manufacture of soft contact lenses. Although PHEMA homopolymer was hydrophilic and has a high degree of hydration, it is not water soluble [50]. Poly(2-hydroxyethyl methacrylate) (PHEMA) is a biocompatible polymer, which is sensitive to pH and temperature and swells in water to form hydrogel. It has been used in dentistry, ophthalmic, scaffold, drug delivery and neutral tissue engineering [51].

Kharismadewi et al. [51] have synthesised graphene oxide-poly(2-hydroxyethyl methacrylate) composite (GO-PHEMA) by dispersion polymerisation in supercritical CO_2 and used for the removal of methylene blue (MB) in aqueous solutions. For this purpose, graphite was oxidised to GO using the modified Hummer's method to have the functional groups of epoxides, hydroxyls and carboxylic acids. The epoxide and hydroxyl groups were located on the GO surface, whereas carboxylic acid was located near the edges. From the –OH group, modification occurred, where silane coupling agent, 3-(trimethoxysilyl)propyl methacrylate (MPTMS), reacted to produce the vinyl groups as the tail of GO. After modification with MPTMS, GO-MPTMS was polymerised with HEMA monomer by dispersion polymerisation in supercritical CO_2 using 2,2′-azobisisobutyronitrile as the initiator. The GO-PHEMA composite as an adsorbent formed a gel in an aqueous solution that provides benefit in the separation of adsorbent. The FTIR, TGA, SEM and XPS were used to confirm the existence of the grafted polymer onto the GO. The composite exhibited an improved thermal loss temperature and a decelerated decomposition rate than GO and GO-MPTMS. The adsorption behaviour of the adsorbent was investigated for the removal of MB from the aqueous solution as a function of the adsorbent concentration, contact time, pH and initial dye concentration. It was evaluated that 99.8% of MB was removed by the adsorbent under the optimum conditions (adsorbent dosage of 15 mg at pH 7, a contact time of 45 min and an initial dye concentration of 10 mg L^{-1} at 25°C). The stability investigation showed that the composite still can remove 81.9% of MB from the solution after six cycles. The adsorption isotherm parameters were fitted well with the Freundlich adsorption isotherm owing a R^2 value of 0.975 and the maximum predicted adsorption capacities of 39.41 mg g^{-1} at 25°C. The adsorption kinetic studies showed that the adsorption behaviour followed a pseudo-second-order model. Furthermore, the thermodynamic studies showed that the adsorption process of MB onto the adsorbent was spontaneous and endothermic in nature with the highest adsorption efficiency observed at 45°C.

Denizli et al. [52] have synthesised poly(hydroxyethyl methacrylate-N-methacryloyl-(L)-glutamic acid) [poly(HEMA-MAGA)] beads by suspension polymerisation for the removal of lead ions. Copolymer MAGA acted as the metal-complexing ligand and polymerised with HEMA. Therefore, the need for the leakage of metal complexing ligand was not necessary. The adsorbent was characterised with FTIR and ^{1}H NMR. The surface area of the poly(HEMA-MAGA) beads was measured by the BET method. The water uptake ratio, elemental analysis and surface morphology of the beads were also evaluated. Batch-type adsorption was used for the removal of Pb(II) ions, and the effect of pH (2.0–7.0) and initial ion concentration (10–500 mg L^{-1}) were determined. Sixty minutes was found to be enough to reach equilibrium. The adsorption of Pb(II) ions increased with increasing pH and reached to a plateau value at a pH of 5.0. It was evaluated that incorporation of MAGA into the polymer structure increased the lead adsorption capacity, ca. tenfold. The adsorption capacities of the poly(HEMA-MAGA) beads for Cd(II), Pb(II) and Hg(II) from synthetic waste water were determined, and the affinity order of adsorption was found as Cd(II) > Pb(II) > Hg(II). The consecutive adsorption and elution cycles indicated the feasibility of repeated use meaning that the newly synthesised beads have a great potential for industrial heavy metal removal applications.

2.7. Polyacrylonitrile (PAN)

Polymers based on acrylonitrile (AN) are easy to prepare and present a reactive pendant cyano group, which can be modified by introducing new functional groups by nucleophilic addition and cycloaddition reactions. In general, AN polymers and copolymers for adsorbent are synthesised through suspension polymerisation to give spherical beads which can be readily handled. The porous-chelating resins can adsorb metal ions more efficiently because the existence of pores would provide convenient diffusion channels for metal ions into the interior of the resins when they are used in adsorption of metal ions in aqueous solution [53].

Godjevargova et al. [54] have synthesised five types of adsorbents from polyacrylonitrile (PAN) by varying its concentration and the composition of the coagulation bath, aiming to achieve different porous structures in the modified adsorbents. The SEM analysis was applied to identify the porous structure of the adsorbents, and specific area, pore volume and pore radius were determined using a porosimeter. The sodium hydroxide and hydroxylamine were used to modify the adsorbents in order to introduce amidoxime and carboxylic groups into the adsorbent. The ability of the modified adsorbents through Cu(II), Ni(II), Zn(II) and Pb(II) ions was determined, and the highest adsorption with respect to all metal ions was achieved at pH values of 3–4. In addition, a rapid (20 min) metal uptake took place in all cases studied. The order of metal uptake for all types of PAN adsorbents was determined as Cu(II) > Ni(II) > Zn(II) > Pb(II). The EDTA was found to desorb the metal ions from the adsorbents. As a result, it was found that the synthesised adsorbents have good adsorption characteristics and are good candidates for the removal of Cu(II), Ni(II), Zn(II) and Pb (II) ions from aqueous solution.

Amine-containing resins based on PAN were synthesised through chemical modifications of PAN with diethylene triamine (DETA) by Kiani et al. [55]. The modified PAN resins (PAN-DETA50 and PAN-DETA150) were used as adsorbent for the removal of Hg(II), Fe(III), Pb(II), Ag(I) and Zn(II) from aqueous solution. The FTIR, SEM and TGA were used to study the surface morphology of the adsorbents. Moreover, the amine content of the resins was determined by titration. The TGA and FTIR spectroscopy study on metal-resin complexes showed the presence of metal ions in PAN-DETA complexes. The SEM micrographs clarified that there are no drastic differences in morphology of resins when compared with unmodified PAN (except PAN-DETA-Zn). The adsorption behaviour of the resins studied for all metal ions confirmed that they have acceptable adsorption capacities. In addition, the adsorption of metal ions was examined under different pH conditions, and it was seen that adsorption was increased with increasing pH values [55].

3. Conclusions

The rise in industrial, medical, technological and domestic activities has led to huge increases in the levels of heavy metals which have serious impacts on both human health and environment. Therefore, the demand for reducing the effects of potential pollutants has increased throughout the world by improving existing techniques and introducing novel adsorbent materials. The preparation of such materials which can remove toxic heavy metals from aqueous media is still a continuing objective of environmental remediation efforts, particularly where there is a need to enhance the adsorption capacity and metal ion specificity. Recently, researchers have focused on polymeric adsorbents for separation and purification purposes due to their valuable advantages such as relatively easy regeneration, perfect mechanical rigidity, vast surface area, adjustable surface chemistry, simple processing and possibility to shape them into most suitable form (e.g. sheets, beads and membranes). This chapter highlighted the practical applications of acrylic-based polymers for the removal of heavy metals ions in water. In summary, it was clearly seen that there is still a need to develop new adsorbents for the removal of heavy metals to introduce simple, rapid and quantitative methods that can be applied in environmental remediation studies.

Author details

Aslı Erdem Yayayürük

Address all correspondence to: aslierdem30@hotmail.com

Ege University, İzmir, Turkey

References

[1] Gupta VK, Gupta M, Sharma S. Process development for the removal of lead and chromium from aqueous solution using red mud—An aluminum industry waste. Water Research. 2001;**35**(5):1125-1134. DOI: 10.1016/S0043-1354(00)00389-4

[2] Hegazi HA. Removal of heavy metals from wastewater using agricultural and industrial wastes as adsorbents. HBRC Journal. 2013;**9**:276-282. DOI: 10.1016/j.hbrcj.2013.08.004

[3] Barakat MA. New trends in removing heavy metals from industrial wastewater. Arabian Journal of Chemistry. 2011;**4**:361-377. DOI: 10.1016/j.arabjc.2010.07.019

[4] Fu F, Wang Q. Removal of heavy metal ions from wastewaters: A review. Journal of Environmental Management. 2011;**92**:407-418. DOI: 10.1016/j.jenvman.2010.11.011

[5] O'Connell DW, Birkinshaw C, O'Dwyer TF. Heavy metal adsorbents prepared from the modification of cellulose: A review. Bioresource Technology. 2008;**99**:6709-6724. DOI: 10.1016/j.biortech.2008.01.036

[6] Shukla A, Zhang YH, Dubey P, Margrave JL, Shukla SS. The role of sawdust in the removal of unwanted materials from water. Journal of Hazardous Materials. 2002;**95**(1-2): 137-152. DOI: 10.1016/S0304-3894(02)00089-4

[7] Charerntanyarak L. Heavy metals removal by chemical coagulation and precipitation. Water Science and Technology. 1999;**39**:135-138. DOI: 10.1016/S0273-1223(99)00304-2

[8] Rubio J, Souza ML, Smith RW. Overview of flotation as a wastewater treatment technique. Minerals Engineering. 2002;**15**(3):139-155. DOI: 10.1016/S0892-6875(01)00216-3

[9] Chen GH. Electrochemical technologies in wastewater treatment. Separation and Purification Technology. 2004;**38**:11-41. DOI: 10.1016/j.seppur.2003.10.006

[10] Babel S, Kurniawan TA. Low-cost adsorbents for heavy metals uptake from contaminated water: A review. Journal of Hazardous Materials. 2003;**97**:219-243. DOI: 10.1016/S0304-3894(02)00263-7

[11] Reddy BSR, Kulasekaran S, Reddy CR, Rajadurai S, Joseph KT, Santappa M. Treatment and disposal of tannery effluents by polymeric scavengers. Leather Science. 1977;**24**:378-383

[12] Choung JH, Lee YW, Choi DK. Adsorption equilibria of toluene on polymeric adsorbents. Journal of Chemical Engineering Data. 2001;**46**(4):954-958. DOI: 10.1021/je000282i

[13] Zhang X, Lin L, Zhang T, Liu H, Zhang X. Catalytic dehydration of lactic acid to acrylic acid over modified ZSM-5 catalysts. Chemical Engineering Journal. 2016;**284**:934-941. DOI: 10.1016/j.cej.2015.09.03

[14] Nasefa MM, Hegazy EA. Preparation and applications of ion exchange membranes by radiation-induced graft copolymerization of polar monomers onto non-polar films. Progress in Polymer Science. 2004;**29**:499-561. DOI: 10.1016/j.progpolymsci.2004.01.003

[15] Benamer S, Mahlous M, Tahtat D, Nacer-Khodja A, Arabi M, Lounici H, Mameri N. Radiation synthesis of chitosan beads grafted with acrylic acid for metal ions sorption. Radiation Physics and Chemistry. 2011;**80**:1391-1397. DOI: 10.1016/j.radphyschem. 2011.06.013

[16] González-Gómez R, Ortega A, Lazo LM, Burillo G. Retention of heavy metal ions on comb-type hydrogels based on acrylic acid and 4-vinylpyridine synthesized by gamma radiation. Radiation Physics and Chemistry. 2014;**102**:117-123. DOI: 10.1016/j. radphyschem.2014.04.026

[17] El-Mohdy HLA, Hegazy EA, El-Nesr EM, El-Wahab MA. Metal sorption behavior of poly(N-vinyl-2-pyrrolidone)/(acrylic acid-co-styrene) hydrogels synthesized by gamma radiation. Journal of Environmental Chemical Engineering. 2013;**1**:328-338. DOI: 10.1016/j.jece.2013.05.013

[18] Wang B, Wang YP, Zhou P, Liu ZQ, Luo SZ, Chu W, Guo Z. Formation of poly(acrylic acid)/alumina composite via in situ polymerization of acrylic acid adsorbed within oxide pores. Colloids and Surfaces A: Physicochemical and Engineering Aspects. 2017;**514**:168-177. DOI: 10.1016/j.colsurfa.2016.11.051

[19] Sezgin N, Balkaya N. Adsorption of heavy metals from industrial wastewater by using polyacrylic acid hydrogel. Desalination and Water Treatment. 2016;**57**:2466-2480. DOI: 10.1080/19443994.2015.1030453

[20] Mouayad QAA, Nagam SAA, Ahmed MAA. Evaluation of poly acrylic acid(PAA) hydrogel beads as adsorbent for the removal of lead(II) ion from water. Journal of Al-Nahrain University. 2013;**16**(3):30-39

[21] Marcelo GV, Antonio GBP, André RF, Antônio CNA, Francisco HAR. Methylene blue adsorption on chitosan-g-poly(acrylic acid)/rice husk ash superabsorbent composite: Kinetics, equilibrium, and thermodynamics. Water, Air, & Soil Pollution. 2017;**228**:1-14. DOI: 10.1007/s11270-016-3185-4

[22] Liu Y, Wang W, Wang A. Adsorption of lead ions from aqueous solution by using carboxymethyl cellulose-*g*-poly (acrylic acid)/attapulgite hydrogel composites. Desalination. 2010;**259**(15):258-264. DOI: 10.1016/j.desal.2010.03.039

[23] Rafiei HR, Shirvani M, Ogunseitan OA. Removal of lead from aqueous solutions by a poly(acrylic acid)/bentonite nanocomposite. Applied Water Science. 2016;**6**:331-338. DOI: 10.1007/s13201-014-0228-0

[24] Rafiei HR, Shirvani M, Ogunseitan OA. Kinetics and thermodynamics of Pb sorption onto bentonite and poly(acrylic acid)/bentonite hybrid sorbent. Desalination and Water Treatment. 2016;**57**:22467-22479. DOI: 10.1080/19443994.2015.1130655

[25] Shirsath SR, Patil AP, Patil R, Naik JB, Gogate PR, Sonawane SH. Removal of brilliant green from wastewater using conventional and ultrasonically prepared poly(acrylic acid) hydrogel loaded with kaolin clay: A comparative study. Ultrasonics Sonochemistry. 2013;**20**:914-923. DOI: 10.1016/j.ultsonch.2012.11.010

[26] Chen H, Li J, Shao D, Ren X, Wang X. Poly(acrylic acid) grafted multiwall carbon nanotubes by plasma techniques for Co(II) removal from aqueous solution. Chemical Engineering Journal. 2012;**210**:475-481. DOI: 10.1016/j.cej.2012.08.082

[27] Yang S, Li J, Shao D, Hu J, Wang X. Adsorption of Ni(II) on oxidized multi-walled carbon nanotubes: Effect of contact time, pH, foreign ions and PAA. Journal of Hazardous Materials. 2009;**166**:109-116. DOI: 10.1016/j.jhazmat.2008.11.003

[28] Manju GN, Krishnan KA, Vinod VP, Anirudhan TS. An investigation into the sorption of heavy metals from wastewaters by polyacrylamide-grafted iron(III) oxide. Journal of Hazardous Materials. 2002;**91**:221-228. DOI: 10.1016/S0304-3894(01)00392-2

[29] Liu P, Guo J. Polyacrylamide grafted attapulgite (PAM-ATP) via surface-initiated atom transfer radical polymerization (SI-ATRP) for removal of Hg(II) ion and dyes. Colloids and Surfaces A: Physicochemical Engineering Aspects. 2006;**282-283**:498-503. DOI: 10.1016/j.colsurfa.2006.02.052

[30] Kaşgöz H, Özgümüş S, Orbay M. Modified polyacrylamide hydrogels and their application in removal of heavy metal ions. Polymer. 2003;**44**:1785-1793. DOI: 10.1016/S0032-3861(03)00033-8

[31] Zhao F, Tang WZ, Zhao D, Meng Y, Yin D, Sillanpää M. Adsorption kinetics, isotherms and mechanisms of Cd(II), Pb(II), Co(II) and Ni(II) by a modified magnetic polyacrylamide microcomposite adsorbent. Journal of Water Process Engineering. 2014;**4**:47-57. DOI: 10.1016/j.jwpe.2014.09.003

[32] Yavuz E, Senkal BF, Bicak N, Poly(acrylamide) grafts on spherical polyvinyl pyridine resin for removal of mercury from aqueous solutions. Reactive & Functional Polymers. 2005;**65**:121-125. DOI: 10.1016/j.reactfunctpolym.2004.12.006

[33] Yayayürük O, Henden E, Bicak N, Determination of mercury(II) in the presence of methylmercury after preconcentration using poly(acrylamide) grafted onto cross-linked poly(4-vinyl pyridine): Application to mercury speciation. Analytical Sciences. 2011;**27**(8):833-838. DOI: 10.2116/analsci.27.833

[34] Suárez E, Paredes B, Rubiera F, Rendueles M, Villa-Garcia MA, Diaz JM. Functionalized glycidyl methacrylate based polymers as stationary phases for protein retention. Separation and Purification Technology. 2002;**27**:1-10. DOI: 10.1016/S1383-5866(01)00163-0

[35] Nastasovic AB, Ekmescic BM, Sandic ZP, Ran–delovic DV, Mozetic M, Vesel A, Onjia AE. Mechanism of Cu(II), Cd(II) and Pb(II) ions sorption from aqueous solutions by macroporous poly(glycidyl methacrylate-co-ethylene glycol dimethacrylate). Applied Surface Science. 2016;**385**:605-615. DOI: 10.1016/j.apsusc.2016.05.165

[36] Nastasovic A, Sandic Z, Surucic L, Maksin D, Jakovljevic D, Onjia A. Kinetics of hexavalent chromium sorption on amino-functionalized macroporous glycidyl methacrylate copolymer. Journal of Hazardous Materials. 2009;**171**:153-159. DOI: 10.1016/j.jhazmat.2009.05.116

[37] Elwakeel KZ, Guibal E. Potential use of magnetic glycidyl methacrylate resin as a mercury sorbent: From basic study to the application to wastewater treatment. Journal of Environmental Chemical Engineering. 2016;**4**:3632-3645. DOI: 10.1016/j.jece.2016.08.001

[38] Şenkal BF, Biçak N. Glycidyl methacrylate based polymer resins with diethylene triamine tetra acetic acid functions for efficient removal of Ca(II) and Mg(II). Reactive and Functional Polymers. 2001;**49**(2):151-157. DOI: 10.1016/s1381-5148(01)00075-x

[39] Yayayürük O, Yayayürük AE, Removal of Cu(II) from water samples using glycidyl methacrylate-based polymer functionalized with diethylenetriamine tetraacetic acid: Investigation of adsorption characteristics. Water, Air & Soil Pollution. 2016;**227**:244-256. DOI: 10.1007/s11270-016-2943-7

[40] Yayayürük O, Yayayürük AE, Karagöz B, Bıçak N. Preconcentration of mercury(II) from waters using glycidyl methacrylate-methyl methacrylate-divinyl benzene terpolymer with diethylenetriamine tetra acetic acid functions prior to inductively coupled plasma mass spectrometric determination. Desalination and Water Treatment. 2016;**57**:27834-27842. DOI: 10.1080/19443994.2016.1178176

[41] Cifçi C, Durmaz O. Removal of heavy metal ions from aqueous solutions by poly(methyl m ethacrylate-co-ethyl acrylate) and poly(methyl methacrylate-co-buthyl m ethacrylate) membranes. Desalination and Water Treatment. 2011;**28**:255-259. DOI: 10/5004/dwt.2011.2174

[42] Al-Muhtaseb AH, Ibrahim KA, Albadarin AB, Ali-khashman O, Walker GM, Ahmad MNM. Remediation of phenol-contaminated water by adsorption using poly(methyl methacrylate) (PMMA). Chemical Engineering Journal. 2011;**168**:691-699. DOI: 10.1016/j.cej.2011.01.057

[43] Salisu A, Sanagi MM, Naim AA, Ibrahim WAW, Karim KJA. Removal of lead ions from aqueous solutions using sodium alginate-graft-poly(methyl methacrylate) beads. Desalination and Water Treatment. 2016;**57**:15353-15361. DOI: 10.1080/19443994.2015.1071685

[44] Mittal A, Ahmad R, Hasan I. Poly (methyl methacrylate)-grafted alginate/Fe3O4 nanocomposite: synthesis and its application for the removal of heavy metal ions. Desalination and Water Treatment. 2016;**57**:19820-19833. DOI: 10.1080/19443994.2015.1104726

[45] Duru PE, Bektaş S, Genç Ö, Patir S, Denizli A. Adsorption of heavy metal ions on poly(ethylene imine)-immobilized poly(methyl methacrylate) microspheres. Journal of Applied Polymer Science. 2001;**81**:197-205. DOI: 10.1002/app.1429

[46] Anbinder P, Macchi C, Amalvy J, Somoza A. Chitosan-graft-poly(n-butyl acrylate) copolymer: Synthesis and characterization of a natural/synthetic hybrid material. Carbohydrate Polymers. 2016;**145**:86-94. DOI: 10.1016/j.carbpol.2016.02.072

[47] Kumar ASK, Kumar CU, Rajesh V, Rajesh N. Microwave assisted preparation of n-butylacrylate grafted chitosanand its application for Cr(VI) adsorption. International Journal of Biological Macromolecules. 2014;**66**:135-143. DOI: 10.1016/j.ijbiomac.2014.02.007

[48] Nithya R, Gomathi T, Sudha PN, Venkatesan J, Anil S, Kim S. Removal of Cr(VI) from aqueous solution using chitosan-g-poly(butylacrylate)/silica gel nanocomposite. International Journal of Biological Macromolecules. 2016;**87**:545-554. DOI: 10.1016/j.ijbiomac.2016.02.076

[49] Nithya R, Sudha PN. Removal of heavy metals from tannery effluent using chitosan-g-poly(butylacrylate)/bentonite nanocomposite as an adsorbent. Textiles and Clothing Sustainability. 2016;**2**(7):1-8. DOI: 10.1186/s40689-016-0018-1

[50] Robinson KL, Khan MA, de Paz Banesez MV, Wang XS, Armes SP. Controlled polymerization of 2-hydroxyethyl methacrylate by ATRP at ambient temperature. Macromolecules. 2001;**34**:3155-3158. DOI: 10.1021/ma0019611

[51] Kharismadewi D, Haldorai Y, Nguyen VH, Tuma D, Shim JJ. Synthesis of graphene oxide-poly(2-hydroxyethyl methacrylate) composite by dispersion polymerization in supercritical CO2: Adsorption behavior for the removal of organic dye. Composite Interfaces. 2016;**23**(7):719-739. DOI: 10.1080/09276440.2016.1169707

[52] Denizli A, Garipcan B, Karabakan A, Senöz H. Synthesis and characterization of poly(hydroxyethyl methacrylate-N-methacryloyl-(l)-glutamic acid) copolymer beads for removal of lead ions. Materials Science and Engineering C. 2005;**25**:448-454. DOI: 10.1016/j.msec.2004.12.001

[53] Liu X, Chen H, Wang C, Qu R, Ji C, Sun C, Zhang Y. Synthesis of porous acrylonitrile/methyl acrylate copolymer beads by suspended emulsion polymerization and their adsorption properties after amidoximation. Journal of Hazardous Materials. 2010;**175**:1014-1021. DOI: 10.1016/j.jhazmat.2009.10.111

[54] Godjevargova T, Simeonova A, Dimov A. Adsorption of heavy metal ions from aqueous solutions by porous polyacrylonitrile beads. Journal of Applied Polymer Science. 2002;**83**:3036-3044. DOI: 10.1002/app.2334

[55] Kiani GR, Sheikhloie H, Arsalani N. Heavy metal ion removal from aqueous solutions by functionalized polyacrylonitrile. Desalination. 2011;**269**:266-270. DOI: 10.1016/j.desal.2010.11.012